PRACTICAL GOATKEEPING

J BARNES

Beech Publishing House
Station Yard
Elsted Marsh
MIDHURST
West Sussex GU29 0JT

ISBN 1-85736-185-7
First published 1977
Second Edition 1982
Third Edition 2004

Originally published as
Exhibition & Practical Goatkeeping
Joan Shields

British Library Cataloguing-in-Publication Data
A catalogue record for this book is available from the British Library.

Beech Publishing House
Station Yard
Elsted Marsh
MIDHURST
West Sussex GU29 0JT

Printed and bound by Antony Rowe Ltd, Eastbourne

Smallholders' Practical Series

OTHER TITLES

Practical Poultry Keeping

Keeping Turkeys

Keeping Guinea Fowl

Commercial Pigeon Keeping

Garden Poultry Keeping

Concise History of the Shire Horse

Keeping Pigs

Farm Stock of Old

Harness Horse

Dutch Rabbits

Old & Rare Breeds of Cattle

Hackney Horse

Practical Fereting

Practical Trapping

Ferrets & Ferreting

Bantams & Small Poultry

Artificial Incubation & Rearing

International Poultry Standards

We have an extensive range of poultry books, covering many separate breeds.

Contents

ACKNOWLEDGEMENTS

The Author wishes to thank the Northumbrian County Council for information on the College Valley Feral goats; the British Museum of Natural History for allowing the use of the paper on the historical background of the domestication of animals by Dr Juliet Clutton-Brock.

— Thanks are also due to the Golden Guernsey Societies of England and Guernsey, and to Marcia Marsden of Sark.

— To the greatest Judge and historian that the Goat fancy has ever had, Mr H. E. Jeffery, whose writings have been perused for dates.

— To the veterinary surgeon who looks after the Author's herd, and 'vetted' the chapter on illness.

— To Mrs Anne May, for the chapter on cheese-making (of which one may truthfully say "the proof of the pudding").

— To Mrs D. Morgan, for illustrating the poisonous plants.

— Mr Briggs for the Feral Cheviot goat photographs.

— And to her husband, L. W. Foster, who drew the plans.

List of Illustrations

COLOUR ILLUSTRATIONS

British Alpine Champion ***Uffdown Oriana*** **(Miss Sadler)**

Toggenberg RM 36 ***Chessetts Cascade,*** **BR. CH (Miss Goldsmith)**

Toggenberg Breed Champion ***Spean Baindidn***

PLATE 1 Champion Goats

Various Cheeses from Goat's Milk

Golden Guernsey *Duvaux Sarnia* GG. 58
Bred by Miss K O Dennis. Owner Miss R Karney

PLATE 2 Goats Cheeses & Golden Guernsey Goat

ACKNOWLEDGEMENTS

Our thanks are offered to those who have given assistance for the new edition. Mrs Charlesworth of Smallholder Publications, Stoke Ferry, Kings Lynn, Norfolk and the author Mrs Margaret Thompson gave permission to draw on the book ***Pygmy Goats*.**

The goat societies:

M/s Sue Knowles
British Goat Society
34-36 Fore Street
Bovey Tracey
Newton Abbot TQ13 9AD

Mrs Margaret Thompson
Pygmy Goat Club
Solomons Farm
Latchley
Gunnislake
Cornwall PL18 9AX

JB ***Elsted Marsh*** ***May, 2004***

PREFACE

WHY KEEP GOATS?

A cow is easier to fence in, her milk is easier to separate without machinery and she gives more.

On the goat's side there are several factors which outweigh the former attributes.

The goat can be kept in with an electric fencer which, being completely portable, can be moved and the animals strip-grazed. Kept thus, three goats will not consume as much grass as one cow.

SAFE MILK

The milk of a healthy goat does not need heat treating before feeding to infants or invalids; it is free from harmful bacteria, is naturally homogenised and, having such tiny fat globules, is easily digested. It keeps very well and can be used for all the purposes to which cow's milk is put.

HOUSING

A cow requires a large shed or portable milking bail, is splashy in her habits and needs cleaning out daily. Two goats can be kept in a shed the size of a standard mini garage (with some small adaptions made by any handy-man). Their bedding can be kept on the deep litter system—thus making cleaning out daily unnecessary.

LONG LACTATION

One cow must have 6–8 weeks dry each year before calving. Two goats, one of which is left unmated each year, will provide a constant yield of milk, for goats of good breeding will milk for two or more years, without freshly kidding.

SIZE

Finally, the smaller size and lighter weight of everything pertaining to goats makes them much easier for the female sex, or the very young, to handle. A reliable 10-year-old child can do most of the work of looking after a couple of Dairy Goats, and the Author's own children were well able to do this at an earlier age.

SHOWING GOATS

The foregoing is purely the practical side reviewed briefly, the hobby of exhibiting is one which would appeal to anyone keen on breeding, rearing and showing livestock. It is also a hobby which can pay its way, if tackled along the right lines from the start.

As with every other form of livestock breeding, a study must be made of blood-lines available which have proved successful, and in the case of the Dairy Goat, this means obtaining the British Goat Society's Herd Books, and having chosen the breed to work with, examining the records, not only of show wins, but also milk records, butter-fat averages, and so on.

VISITING AGRICULTURAL SHOWS AND GOAT CLUB SHOWS

At all the bigger Agricultural shows there is a Dairy Goat section which classifies all the breeds, although not always separately. Here you may see the best specimens of those breeds, in good condition and at all stages of growth, from two-month-old kids to mature milkers. Normally, you will not see the male of the species, for he has a separate show, to which the public are not usually invited on account of his forceful odour. This he wears most strongly during the rutting or breeding season, which is from September until about mid March, but an adult male goat can be smelt some distance away even in summer and so the poor chap is segregated.

FEMALE GOATS DO NOT SMELL

It is of course the male's characteristic odour which first caused the rumour that "Goats *smell*" to be put about. Fortunately, those household goatkeepers who are not serious breeders and do not intend to run to a herd, do not need to keep a male to serve their couple of household goats once a year for the local goat club can supply a Stud Goat List, bearing the names of pedigree males in all breeds within reasonable reach; some of these males are mobile and can be brought round by appointment

LOCAL GOAT CLUBS

Local clubs exist in many areas and the members are only too pleased to give advice on matters relating to keeping goats. They may also assist in obtaining stock and getting established. When these clubs are affiliated to the British Goat Society a list of addresses may be obtained. The annual fees are quite low so it is worthwhile being a member, as well as being a member of The British Goat Society.

A DELIGHTFUL PET

So much for the Dairy Goat as a household provider and as a small livestock breeding proposition to add interest for the owner who likes to exhibit. Some breeds of Dairy Goat in particular, and more than others, make the most amusing and affectionate pets, and although the Author deplores those who seek to find a young kid 'to be a pet for the children and keep the grass down' as their reasons, the fact remains that the happy well-cared for pair of household goats *will* make pleasant pets into the bargain.

NO TETHERING—BUY A GOOSE OR A LAWN MOWER

Let no one who intends to tether their goats on a small scrubby patch of poor land read further, unless he or she is prepared to learn to become a good goat husbandman! The Author's intention and ardent hope is to show the right way to keep a well-bred contented animal in profit, for a long and useful life at the least possible expense, with a probable reward in the shape of show wins and the profitable sale of surplus young stock, as well as plenty of delicious home-produced dairy produce.

Old Style Nubian Goats

CHAPTER 1

THE ANGLO-NUBIAN BREED

ORIGINS

Our Anglo-Nubians are descended from the first Eastern goats to arrive in England about 1850, from the indigenous* English goats (no longer registered) and a variety of others, probably including some surreptitious imports made after the restrictions on bringing livestock into the country were imposed.

Two so-called Nubians were also imported from Paris in 1883, but none of these early goats were recognised as pure bred and they were only used as foundation females with the pure Nubian males, imported as such, later.

Early goat breeders went down to the London Docks and eagerly bought up any foreign goats which came ashore from liners, which of course in those days lacked any refrigeration. The goats were on board to give a supply of

* *See* Chapter 7. Strictly speaking there are no *indigenous* English goats, but native long enough to be regarded as such.

fresh milk for young babies returning with their parents from the East.

The first recognised pure-bred Nubian goat to be imported was registered in the new Herd Book as Nubian **number one.** He was called *Sedgemere Chancellor* and was owned by Mr Sam Woodiwiss, of Bulldog fame. Mr Woodiwiss purchased *Chancellor* from the well-known animal importer, Mr Jamrack, in 1896.

Nubian **number two** in the Herd Book was *Sedgemere Sanger,* a Zaraiby male imported in 1904, and in the same year came Nubian **number three,** a Chitral male named *Bricket Cross.*

Nubian **number four,** *Bricket Zoo,* completed the four males which were recognised by the British Goat Society as pure-bred Nubians (though of different districts).

The first naming of our breed as Anglo-Nubians occurs in Volume Two of the Herd Book around 1893, but it was not until 1910 that the British Goat Society opened the Anglo-Nubian Herd Book. Twenty-nine of the progeny of *Sedgemere Chancellor,* and seventy-two of *Bricket Cross*'s, are recorded therein.

Gradually, as in the case of the other evolved breeds, the Anglo-Nubian progressed both in numbers and in breed type, and the present day qualifications for entry were determined.

Until the arrival in England of the two Swiss breeds, the Toggenburg and the Saanen and, soon after, the formation of the larger heavier-milking British Toggenburg and British Saanen, the Anglo-Nubian held its own, but could not compete for milk yield with these new breeds, which had been bred much more scientifically, with emphasis on production records. Whereas the Anglo-Nubian of those early days gave about 9–10 lb of milk at most, the good ones (the crossed goats), were already soaring to show yields of 21 lb, and it is not unusual to find show reports of milking competitions around 1927 where the first five goats gave yields of over 15 lb (1½ gal) in the 24 hours. Butterfats of the early Anglo-Nubians seem to have been steady between 4·5 and 6%.

This set-back in the breed's popularity has now been reversed and the registrations bid fair to outnumber even the British Saanen. This is because the breed has improved so greatly within the last forty years, both in milk yield and in type, largely owing to the skill of the keen breeders;

perhaps the most successful during the writer's time being Mr and Mrs Egerton and Miss Rochford (Malpas and Berkham herds). Many new breeders and exhibitors are now coming forward and are finding the breed lucrative on account of the demand for export.

MILK RECORDS

In 1933 the highest recorded Anglo-Nubian gave 2848 lb in 365 days; the highest yield today stands at 4271 lb. The highest 24-hour show yield is 16·7 lb given by *Spean Kadida* owned by Mrs Partridge. The average over 204 Anglo-Nubians entered in 1974 milking competitions was 7·19 lb with an average of 4·754% butterfat.

This improvement continues, the most recent example being *Champion Redgill Hebe,* owned by Mrs Stevens, who has won eight supreme (inter-breed) championships and seven Inspection/Production Challenge Certificates in one season, becoming the fourth member of this breed to gain her full championship.

A DESCRIPTION

So much for the history of this breed since its inception. What does it look like?

The Anglo-Nubian is now one of the largest and heaviest of our breeds, females weighing up to 200 lb and males up to 300 lb. It is quite unlike any other breed of goat; perhaps the Syrian comes nearest to it, but these goats, although they have very long ears, do not have the very accentuated Roman Nose.

It may be any colour and is often spotted, mottled or marbled, with two or three colours. Markings resembling Swiss breeds are frowned upon. Quite often these goats are red with a black eel stripe and black markings on legs.

The head is shaped rather like that of the camel, with high, slanting, supercilious eyes. The ears should be at least as long as is necessary to reach the nose, and set very low and flat to the skull. Flying ears, or half erect ears, are a fault.

The nose is very pronouncedly Roman, with a comparatively small, narrow mouth, in which one must not expect to find the teeth set as in a Swiss or English goat's mouth. Because of the very different set of the jaw the teeth may be slightly undershot, as they would be in a Peke or a

* **The latest figures should be obtained from the British Goat Society.**

Boxer dog, but they should not be visible when the mouth is closed.

It seems likely that the more accentuated the Eastern breed type the more undershot the jaw is likely to be. This is very noticeable in the old photographs of those early imported Nubians, particularly the Zaraiby goats, which have been said to be the best type.

We have no animals comparable with these today; indeed, it now seems that they would be unacceptable in the show ring, but some latitude over teeth must be allowed if the true Nubian type is to be retained.

The neck should be long and fine, fitting smoothly into distinct withers and fine shoulders. The brisket is well developed and the belly capacious, with a noticeable dairy wedge showing.

The legs are strong and straight, with good bone and rather upright pasterns, and the hock joint shows little curve, being comparable with the straighter stifle of the Chow Chow dog.

Some breeders think that the back line should slope upwards from the shoulders to the rump. The writer prefers a longish, level back line and without a steep slope from rump to tail.

The Anglo-Nubian should from every angle present a wide strong frame and a well attached udder of good spherical shape will prove the practicality of the show type to strive for, in every breed; for without this asset many an old goat with udder "nearly touching the floor" comes to grief before her working days are really over.

It may well be that the long, close ears are an attribute evolved through the centuries for keeping the desert sand out. Certainly the straight stifle is better for shuffling through sand than jumping, which the breed seldom does.

MANAGEMENT

The Anglo-Nubian has been said to be a poor grazer, but the Author has found that they will graze good grass well. It must, however, be treated as a crop and properly maintained if it is to account for maintenance and part of the production ration, and this of course applies to all breeds of goat. But old tufty neglected stuff will not be eaten by the Anglo-Nubian, although they will browse happily upon brambles, nettles, docks and thistles, like other goats.

Figure 1—Champion Redgill Hebe *Q* Br. Ch. Anglo-Nubian*
(owned by Mrs M. Stevens, Sussex)

The usual branches, with kale in winter, are essential for health, and special attention should be paid to the mixing of the concentrate ration, bearing in mind the extra-high butterfat percentage of their milk.

The breed, if unsettled, as for example when parted from a stable companion or when a female is in season, can be rather noisy, having a strange bleat unlike other goats, and what really amounts almost to a growl when annoyed, but normally they are quiet and peaceful, and have a great clan spirit.

GOOD QUALITIES

From the show point of view this is an excellent breed to choose, for a good specimen is so eye-catching and has such presence that it is difficult for any judge to pass her over.

Whilst not the best choice for really Spartan conditions, well-fed and well-housed herds in the North do extremely well.

Anglo-Nubians kids are the most appealing of all goat kids and seem to have a special affinity with humans from a few hours old, snuggling up and quite obviously liking being nursed and cuddled.

It is interesting to note how many are exported to the countries of their origin. Almost all orders for the tropics specify goats of this breed. Recently 28 Anglo-Nubians set out for Khartoum to form a nucleus to improve the milk yields of the Sudanese Nubians.

BREED STANDARD

Head—Facial line convex, ears long and pendulous, set low and lying flat to the head. No tassels.

Colour—Any colour or combination of colours, but ideally not showing facial stripes. Coat short and fine.

Conformation—Tall and upstanding, back straight and level. Hind legs set well apart without pronounced angle at hock. Tail straight and set high.

CHAPTER 2

THE TOGGENBURG

A SMALLER GOAT

The Toggenburg goat is the most delightful, clever and funny little animal of all the breeds. Her smaller size, hardy constitution, attractive colouring and splendid conformation, endear her to everyone who appreciates these qualities, and her lower yield is amply compensated by her ability to live cheaply off the land to a very great extent, and to do well in rather Spartan conditions. The length of lactation in this breed is very good, and although one would not expect much more than 9–10 lb of milk from an animal weighing about 115–125 lb, it is quite usual for the 'Tog' to give ½ gal daily during the winter months, and to increase to two-thirds her former peak yield with the coming of spring, again without kidding.

HISTORY

The first Toggenburgs came from the Canton St Gallen, in Switzerland, via Paris, in 1884, and after the prohibition of importing animals came into force in 1906, no more could be brought in until 1922, when a permit was obtained to import 17 Toggenburgs. Another long gap in new blood made breeding difficult until, in 1965, another six Togs arrived. Unfortunately, although show type and milk yields were considered, little attention seems to have been paid to butterfat content on this occasion, and the boost to yields has not been constructive for the breed on account of the lowered butterfat content of the descendant's milk.

The Toggenburg, like the Saanen, is registered in a 'closed' Herd Book, which contains only those goats which were actually imported and their pure bred descendants. Thus no grading-up into this section of the Herd Book can take place and any cross-bred progeny are entered into another appropiate section, as will be seen in the chapters on the British Breeds. The Toggenburg Section was opened in 1905.

Ninety-six goats qualified for immediate entry, but with the coming of the larger and heavier producing breeds the little Toggenburg has unfortunately taken a back seat in this country, although popular in the USA and in Canada.

The value of the pure breed must not be overlooked, however, for it is vital for keeping true type and good conformation in the British Toggenburg breed. One often sees animals of this breed which need a new infusion of pure blood to improve conformation, and the notable milkers in the British Toggenburg Section, which have one Toggenburg parent or grandparent refute the often-heard remark that "using a Toggenburg male on a British Toggenburg will reduce the milk yield of the progeny".

RECORD YIELD*

The highest officially recorded milk yield for 1975 was given by a British Toggenburg, R50 *Handmaiden Q*1* (Breed Champion), who was sired by the imported-in-dam Toggenburg, *Tanatside Hansel*—she gave 5050 lb of milk in 365 days, at 3·91% BFs. *Handmaiden,* at the age of nine years, has given 17 lb of milk at 3·49 and 3·6% BFs in the milking competition at the Royal Agricultural Show, which she won.

* **The latest figures should be obtained from the British Goat Society.**

Figure 2—Pure Toggenburg Breed Champion
RM35 Spean Corlovme *1, *as a goatling (18 months)*

There have been many others with stupendous yields, some in the Northmoor Herd owned by Mrs Frankland and one bred by the writer, which was a first cross pure Toggenburg/pure Saanen, which gave 5100 lb in 365 days.

BREED RECORD YIELDS *

For some time, the Author's home bred Toggenburg, *Spean Corlovme *1* (Breed Champion), held the breed record with an official yield of 3500 lb in 365 days, but this has now been beaten twice, by *Alderkarr Damaris QX* (Breed Champion), bred by Mrs Hetherington, which gave 3583 lb at 3·82% BFs, and recently by *Pippins Candida QX* (Breed Champion), bred by Mrs Robinson, which gave 4120 lb at 4·01% BFs. There has now been a full Champion female Toggenburg, Champion *Spean Meliflous QX* Breed Champion, whose officially recorded milk yield was 142 in 365 days as a second kidder. Previously only Toggenburg males had qualified.

APPEARANCE

Essentially a neat, active, well-made little goat, with strong hocks, a good deep body, fairly wide chest and pretty head with happy interested expression.

In colour the Toggenburg is a soft shade of fawn varying to mouse or brown, dark chocolate being undesirable. The hair is soft and silky, but may be long all over the goat or long on the flanks and down the spine. These fringes are often bleached-looking, and either silvery or golden, which is most attractive when combed out.

The distinctive white markings, known as "Swiss", are apparent on all three Swiss-type coloured goats: Toggenburgs, British Toggenburgs and British Alpines, and consist of white facial stripes from above the eyes down to the muzzle, around the edges and insides of the ears and down the legs from the knees and hocks, continuing on either side of, and under, the tail.

The ears are small and very pricked and the animal has a remarkably long brisk striding walk.

* **The latest figures should be obtained from the British Goat Society.**

Figure 3—R142 Champion Spean Meliflous *Q*, Breed Champion. The only pure Toggenburg CHAMPION that there has been to date.*

(owned and bred by Joan Shields)

BRITISH GOAT SOCIETY'S BREED STANDARD

Head—Ears erect and pointing forward, facial line slightly dished.

Colour—Fawn, with Swiss markings.

Conformation—Strong but fairly fine bone. Generally smaller than the other breeds.

Old Style Toggenburg

CHAPTER 3

THE SAANEN BREED

ORIGINS

Credit for the development of the Saanen breed must go to those early pioneers in the Saane and Simmental valleys in the Swiss county, or canton, of Berne, where it originated and from whence it has spread almost all over the world. For many years the Saanen has been kept and used in conjunction with other breeds in Germany, France, Holland, Belgium and, during the period between 1890 and 1900, and again in 1903, a few Saanens were brought into England but in insufficient numbers to maintain them permanently as a pure breed.

When a 'Swiss' section of the Herd Book was opened by the British Goat Society, these first few Saanens were numbered S1 to S5, and from 1918 to 1922, when a much more practical importation was made from Holland, there was no real progress with the breed, excepting only as a great stimulus for the indigenous and cross-bred milkers, whose progeny were often described as Anglo-Nubian-Swiss.

In 1922, 11 Saanen male goats and 18 Saanen females were imported from Holland; the Swiss section of the Herd Book then became the Saanen section and the new imports were numbered from S10 to S38.

This proved a real benefit to the goat breeders, both those to whom keeping the Saanen breed pure and the Herd Book closed was of prior interest, and also to those who had already turned their thoughts to the formation of the British Saanen breed.

FAULTS OF THE EARLY IMPORTS

As so often happened with the kids of the other Swiss closed herd book breed, the Toggenburg, too large a percentage of Hornless Saanen kids were infertile, hermaphrodite or inter-sexed, and this is thought to have been accentuated by the habit of breeders in their native land of destroying *all* horned kids at birth in a futile endeavour to produce entirely hornless herds. After Professor Hagedoorn had done considerable work upon the incidence of hermaphroditism in naturally hornless goats, the British Goat Society were persuaded to amend their regulation stating that no disbudded or horned males could be registered, and this in turn alleviated the situation, for Saanen and Toggenburg breeders were quick to take advantage of the opportunity of using one disbudded parent, and it is now accepted that one parent should be disbudded. The chances of a horned hermaphrodite, although this can occur, are about one in one thousand or less.

Another fault in the early Saanens were cow hocks and weak stifles with slack long pasterns, but these faults have largely been eliminated by judicious breeding.

MILK RECORDS *

The average milk yield for the breed during the shows of 1974, taken over 55 goats, was 9·066 lb, with an average butterfat of 3·295%. The highest recorded yield in a 365-day lactation is 6284 lb and the highest 24 hour yield at a show 18 lb 13 oz.

There were 94 pure Saanens registered in 1974.

* **The latest figures should be obtained from the British Goat Society.**

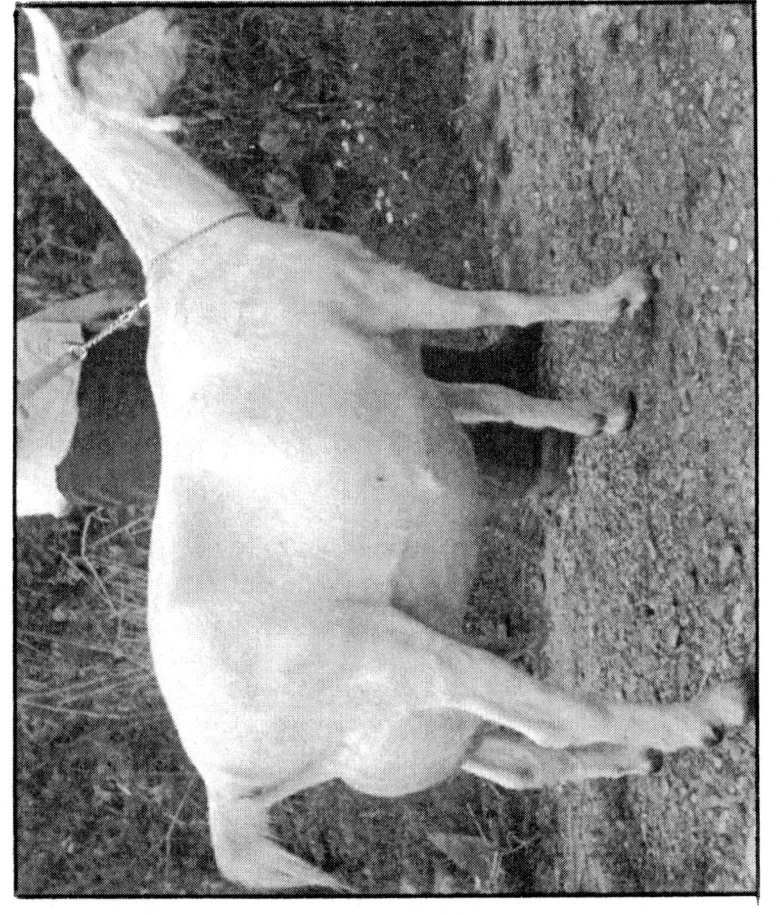

Figure 4—R33 Mostyn Merrymiss, Supreme Saanen Champion. Yield 5050 lb in 574 days

(Bred by Miss J. Mostyn-Owen, owned by Mrs P. Carter)

APPEARANCE

In size, a medium goat weighing about 145–150 lb, with males up to about 250 lb, the Saanen is white, shorter on the leg than the British Saanen, has smallish prick ears and well developed body. The coat is short but may have longer fringes.

In character, the Saanen is quiet, placid and a good doer.

BREED STANDARD

Head—Ears erect and pointing forward. Facial line slightly dished.

Colour—White.

Conformation—Strong but fairly fine bone. Deep body, legs tending to be short in proportion to body.

Old Style British Alpine

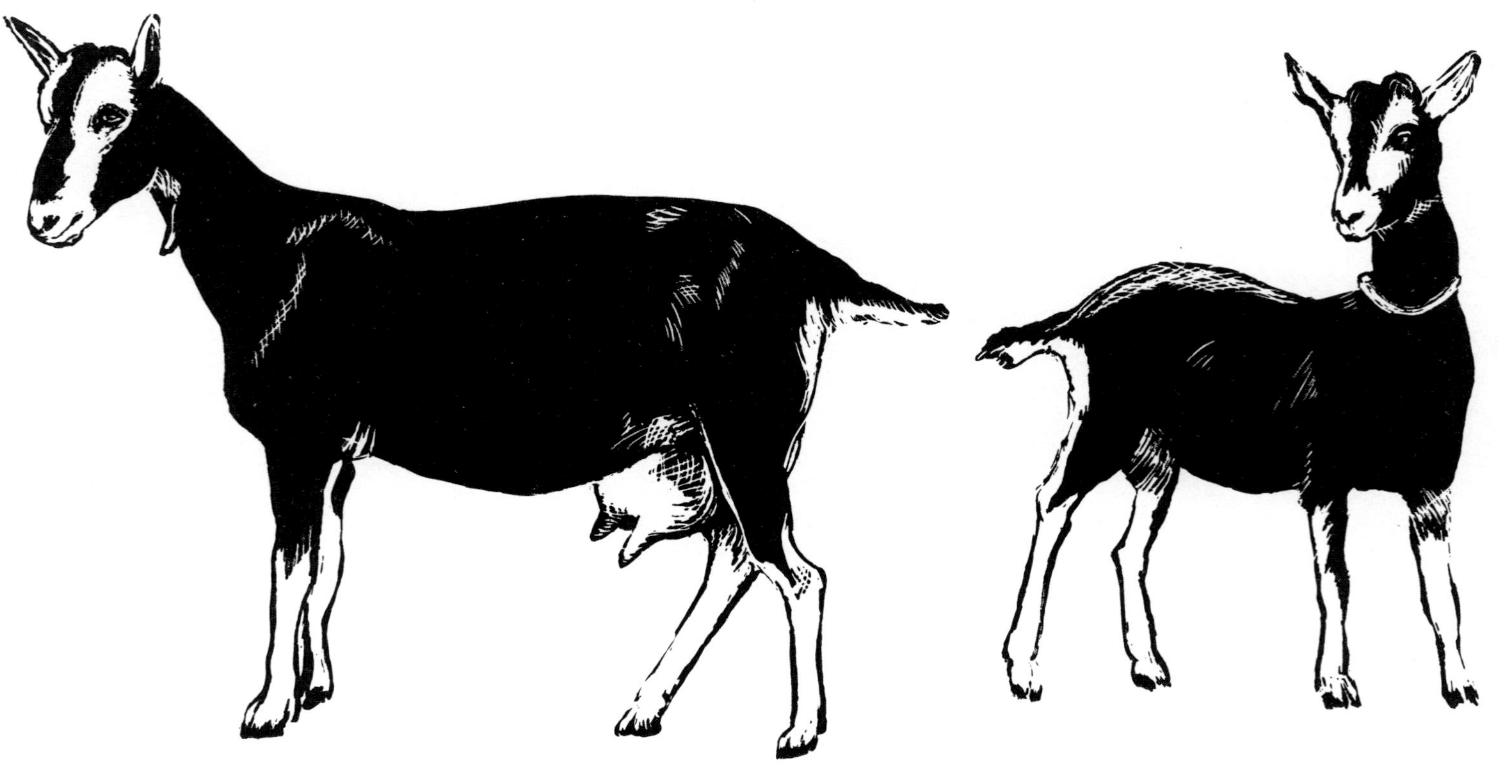

British Alpines

CHAPTER 4

THE BRITISH ALPINE BREED

ORIGINS UNCERTAIN

The British Alpine is one of the most difficult breeds to attempt to establish as a show herd, for the origins of this breed are not so clearly defined that one can say with certainty what any line of breeding will produce. Nor are there sufficient first class show specimens without faults to enable the serious breeder to line-breed, and to hope thus to fix a strain so excellent that no goat in his herd will develop a poor shaped udder with ugly teats. Many lovely goatlings leave the show ring for good on becoming milkers.

That there are black and white Swiss goats with the typical markings found in the Toggenburg seems fairly certain, but it is generally agreed that the British Alpine had as its most prominent forbear the black and white *Sedgemere Faith,* who was correctly Swiss-marked except for a white blaize on the forehead.

THE BRITISH ALPINE HERD BOOK

Black and white Swiss markings seem to have cropped up from the time of *Sedgemere Faith*'s arrival in this country in 1903, when she arrived via the Paris Zoo. In 1920 the British Goat Society had begun to describe these goats as British Alpines and had put on classes for the breed at the bigger shows, and in 1926 the British Alpine Section of the Herd Book was opened.

Qualifications for entry into this section were at first confined to the goat's appearance; she was inspected by a licenced judge and was required to be entered first in the British Section.

By 1935 it was found that entry could be restricted to actual breeding and sufficient British Alpine entries in the pedigree were now required. At this time a Register was also opened so that goats which looked like typical British Alpines could be entered without the percentage of 'pure blood' breeding behind them required for the British Alpine herd book.

The Register continued until 1943, when it was then closed and all British Alpines since that time have qualified on their breeding.

THE FIRST BRITISH ALPINE HERDS

The Pytchley and Didgemere Herds

In 1911 the first famous British Alpine herd emerged, that of the late Mrs Soames; soon afterwards Mrs Abbey's Didgemere herd came to the fore and has consistently produced animals of beautiful type with tremendous milk yields over many years.

As so often proves to be the case, one breeder can lay claim to having founded a breed and brought it to the fore as regards popularity, through consistently showing good specimens. One could safely claim that all the best British Alpines today have Didgemere blood not so very far back in their pedigrees.

MILK YIELD OF THE BREED

The average milk yield of this breed taken on 196 British Alpines at the 1974 shows is 9·33 lb, with average butterfat content of 3·504%. The highest recorded British Alpine gave 3335 lb at 4·05% BFs during the same year. The 24 hour show yield is 23 lb 8 oz.

APPEARANCE AND CHARACTER

Probably one of the most striking looking of the breeds, a good specimen is so shiningly black and dazzlingly white; it is tall and elegant into the bargain.

It is important that the colouring is good—sandy white markings, or less than jet black colouring, is a fault.

The hair should be sleek, smooth and short, but there is sometimes a transitory period when the winter coat is changing and the black markings are rusty, with a longer coat which is giving way to the jet black spring coat. White under the belly is a fault and one which is fast disappearing in this breed, as with the British Toggenburg. This mismarking can spoil the appearance of an otherwise good animal by making it seem shallow in the body.

British Alpines are very good browsers and are probably more content with part woodland and part pasture; if confined on a small acreage they tend to stand looking longingly over the fence, they are terrific jumpers although a well maintained electric fence usually contains them.

A fault which most novices make when starting goatkeeping with this breed lies in not supplying sufficient bulk. It can stunt the kid's growth if, from quite an early age, they are not encouraged to eat all they will of good roughage, as they have a large frame to grow.

BRITISH ALPINE BREED STANDARD

Head—Ears erect and pointing slightly forward.
Colour—Black. Swiss markings.
Conformation—Tending to be rangy.

Herd of Toggenbergs

CHAPTER 5

THE BRITISH TOGGENBURG BREED

A POPULAR GOAT

The British Toggenburg breed is deservedly popular both as a free range goat and for stall-feeding. Larger than the pure Toggenburg (her forbear), she requires more fodder but also gives more milk. Of the same attractive fawn shades as the Toggenburg, with white Swiss markings, a darker chocolate colouring is also permissible. The hair is usually shorter, although near Toggenburg blood often accounts for some long fringes which are not considered a fault.

Generally British Toggenburgs tend to have longer ears, a larger rangier frame and a straighter facial line than the Toggenburg.

ORIGINS

After the British Saanen and the British Alpine goats became recognised breeds with sections of the Herd Book, the breeders who had brought about (through consistently using male Toggenburgs of pure blood with Swiss and English goats) a good type of dairy animal—breeding reasonably true to type—arranged the British Toggenburg Section of the Herd Book. In 1925 this breed was officially started.

At first, as with the other made breeds, entry into the Herd Book was by way of the numbers of animals in a goat's pedigree and she might qualify for the British Toggenburg Register, a jumping-off spot towards the breed section to be entered by her progeny later. This Register was closed in 1943 when it proved to have served its purpose, and the breed was firmly established.

MILK YIELDS

There have been many outstanding milkers in this breed, the average taken over the 1974 milking competitions (218 British Toggenburgs) is 9 lb at 3·503 butterfats. There were, in 1974, 343 British Toggenburgs registered as against 154 in 1945, which clearly shows that the breed is holding its own. The 24 hour show yield is 23 lb 13 oz.

The British Toggenburg could definitely be described as a satisfactory animal for beginners, being of reasonably calm disposition, easy to manage, and having inherited some of the stamina of its pure Toggenburg forbear.

There is a distinct inclination for the present-day British Toggenburgs to lack width behind the shoulder and to appear somewhat shelly—breeders would do well to make an occasional return to the pure Toggenburg, which they are indeed lucky to have recourse to, a facility not afforded to the British Alpine, the Anglo-Nubian or the newer Golden Guernsey breed.

The British Toggenburg, whilst not in such demand for export as the Anglo-Nubian has, nevertheless, reached several foreign countries and has given a good account of herself. The interest shown in the few which did arrive in the USA before the scrapies ban on imports from this country made further shipments practically impossible, clearly showed that here would have been a splendid opening for the breed.

Figure 5—British Toggenburg goatling Balaams Norma. A consistent winner

(Bred and owned by Mr and Mrs Derek Hall)

THE BRITISH GOAT SOCIETY'S BREED STANDARD

Head—Ears erect and pointing slightly forward. Facial line straight or slightly dished.

Colour—May vary from fawn to chocolate. White Swiss markings.

Conformation—Larger and more rangy than the pure Toggenburg.

Note—As with the Saanen and British Saanen, the use of the pure breed on one partner in the British Toggenburg section of the Herd Book results in progeny in the British Toggenburg Section, thus they are interchangeable one way only, the pure breed section being, of course, closed.

Stalls With Slatted Floors

CHAPTER 6

THE BRITISH SAANEN BREED

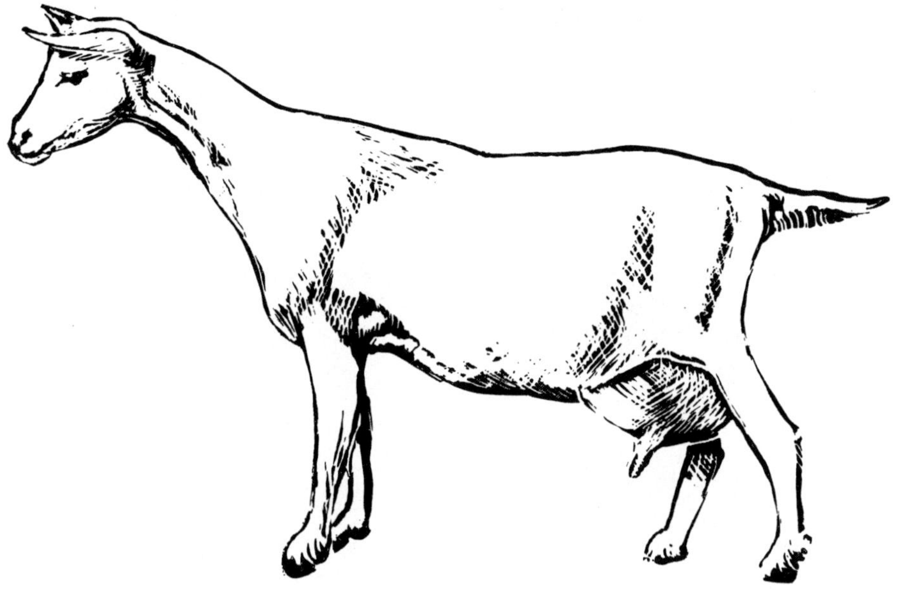

ORIGINS

The British Saanen was until recently without doubt the most popular of the breeds, as registrations showed. When one realises that some fifty years of careful breeding for milk, and size and length of lactation had gone before, with the Anglo-Nubian **x** Swiss goats that had been bred from the original importations, and that these descendants of that small influx of Saanens then proceeded with added vigour and yet more milk from the 1922 importations. Breeders not primarily interested in the closed Herd Book pure Saanens found that they had established a superior large breed which gave even more milk than its parent breed, the Saanen. Most of these good animals were white, this being dominant, and these were known as British Saanens to distinguish them from the Saanens.

In 1925 a section of the Herd Book was opened for the British Saanen. To be entered in this section the goat had to be eligible for the General Section (in which the crossed Anglo-Nubian-Swiss goats were registered), and it also had to be passed on inspection by two licensed British Goat Society judges, after it became six months old, as being of the desired type.

In 1935 a Breed Register was inaugurated in which might be classed those goats almost ready to attain the breed section. By 1943 the breed was well established and the Register was discontinued, so that all entry into the British Saanen Breed Section was by breeding only.

Because of the dominance of white over colours this breed rarely produces wrong colours, although black skin spots are found, particularly on udders, as with the Saanen. The conformation is very much more sound and stable, probably because of the admixture of other breeds having better hocks, stronger stifles and more upright pastern joints.

For the same reason, the pretty dish face of the Saanen has to some extent given way to a straighter facial line.

EARLY BREEDERS

To mention owners' names is to leave out, with certainty, some who played a great part in the early history of this breed, but to mention their prefixes is essential for today's aspiring breeders whose interest leads them on a search through early pedigrees, and in these will be found the names Heddon, Ridgeway of Weald, Didgemere (later of British Alpine fame), Theydon, Broxbourne and Springfield occurring many times. However, the name which above all others has persistently come to mean all that is good in the British Saanen is, of course, **Mostyn,** a prefix which passed from mother to daughter, and under which name some, indeed many, of our best goats of all time have been exhibited and officially recorded. From 1933 or thereabouts to the present time is a long spell for one prefix to have so dominated a breed and to have consistently produced winners in the ring and at the pail. The number of full champions bred and made in this herd is a great credit to the breeders.

It must be added that the pure Saanen is also bred—and has certainly not been eclipsed by her larger relation—although this is not a large herd, and it proves just how much skill has gone into breeding, that from small numbers such quality is consistently bred.

APPEARANCE

The British Saanen is a large, sound, well-made white goat, usually with a well-shaped udder. A good specimen is certainly handsome, if kept clean. A soiled white animal is most unattractive, so particular attention should be paid to the bedding. Once badly stained, a white goat is extremely difficult to get spotless again, no matter how much one scrubs. The weight is in the region of 170 lb.

CHARACTER

One of the British Saanen's greatest assets is her calm, docile character, few specimens showing aggression (although there is one unpleasant strain),. being quiet with other animals and willing to stay within bounds if the grazing is good. She is the ultimate choice for stall-feeding, giving a really good account of herself when wellbred and well fed.

FAULTS

Just occasionally one sees a rusty back-line, shaded cream along the neck, said to come from the import, Betelgeuse, with longer hair.

Another fault is a coarse shoulder, with an ill-defined wither and a sloping rump. A coarse wide face with unfeminine expression and a tendency to a roman nose is not pretty and usually denotes a rather poor milker. Slack pasterns may be caused by an over-fat animal in its youth, but may also be a throw-back to the Saanen forbear, as may the occasional thin weak stifle.

PERFORMANCE

A performance of 6442 lb in 365 days, a recorded lactation yield for this breed with 23 lb 2 oz weighed in a 24-hour period at a recognised show. The average yield taken over the 247 British Saanens in the 1974 milking competitions at shows was 10·62, with average butterfats of 3·47%.

BREED STANDARD

Head—Ears erect and pointing forward.
Colour—White.
Conformation—Strong bone, deep body, legs longer, tending to be more rangy than the pure Saanen.

Old English Stud Goat

CHAPTER 7

THE OLD ENGLISH GOAT AND THE FERAL GOATS OF THE BRITISH ISLES

ORIGINS

These, the Old English Breed and the Feral Goats, are to be dealt with in one chapter because it seems extremely probable that both stem from the same original source.

At the end of the Ice Age, with denser population and climatic changes taking place, there arose a need for various commodities such as grain, meat, bone, skins, fats and oils for lighting.

At the beginning of the Pastoral communities in the Neolithic period, the storage of meat on the hoof by tethering animals superseded the use of hunted animals such as aurochs, wild boar, gazelle, deer, foxes and rodents, fish and molluscs, which had hitherto, together with birds and plants, been the principal foods around 9000 B.C.

THE FIRST GOATS

This is clearly shown by remains in Western Asia. Sites dated 7000 B.C. contained cultivated grain and, for the first time, sheep and goat remains, which outnumbered all other species.

These sheep and goat remains are of smaller size than those in the wild, owing to the restricted grazing imposed by mankind.

Deer, gazelle and other animals could not have been domesticated as easily as sheep and goats, for they do not breed so freely in captivity, nor are they so easily confined. Because Man was becoming a farmer, he had to protect his grain.

As yet, no selective breeding was practised, although some males not required for breeding are thought to have been castrated.

The early Egyptians, however, did make some effort to evolve definite breeds, and are believed to have been pioneers in animal development. In Northern Europe there is no evidence of selective breeding until the Roman period.

THE FIRST FRIEND

Dog, the first animal to be domesticated from the wolf, is carbon-dated much earlier—12000 B.C. This at Palegawra, in North East Iraq, and also in Yorkshire, but there is no trace of goats in the British Isles at this time.

The breeding of domesticated dogs would have meant the end of the wolf packs in the vicinity of human communities and, sadly, this behaviour of Man towards the progenitors of his domestic stock has remained the same throughout the ages. Fortunately, the wholesale slaughter of the Feral goats of the British Isles is at last checked.

SURVIVAL

The two ungulates, sheep and goats were, then, the first food animals to be controlled by the early farmers. The goats, descended from Capra aegagrus, the Scimitar

Horned goat, and the sheep, from The Mouflon, Ovis orientalis. The *former,* still to be found wild in Turkey and upon some Greek Islands, including Crete, and the *latter* in Cyprus, Turkey and Western Asia.

If the prototypes of the Soay sheep were taken to the Western Isles and the Outer Hebrides in prehistoric times and have, since the evacuation of St Kilda, become completely Feral, it is reasonable to suppose that the same may apply to the 'wild' goats of the Cheviots and other mountainous regions of the British Isles. There seems little doubt that they were domesticated by the late Neolithic farmers, probably having been brought here by the Phoenecians, and are Feral.

HORN SHAPE

If this latter is true, it is surprising that our Feral goats do not have the long lop ears of the Asiatic breeds. Their ears are usually small and very pricked.

The scimitar horns do seem to occur quite frequently in Nubian types, and one at least of the four original imported Nubians had scimitar horns.

From time to time attempts have been made to classify different types of goat by horn shape, but in practice this does not work out as twisted horns occur in several types, as do scimitar horns.

The base of the Swiss breeds' horns would seem to be broader than the base of the Asiatic goat's horns, and thus they are sometimes more difficult to disbud successfully.

EARLY GOATS

As noted, the origin of the Old English goat goes right back to the late Neolithic period and no indigenous British goat exists in truth.

For many years the Old English type reigned unchallenged, until the importation of the Nubians and, later, the Swiss breeds, then crosses between these, with admixture of Old English blood became popular.

It was the coming of the Nubians and, later, the Swiss breeds, which caused the end of the Old English Probationer's Record, with their higher yields, and more stylish types. The Old English were used in fashioning the new breeds, together with the Nubians and the Swiss, but as a breed, with a part of the British Goat Society's Herd Book, they ceased to exist in 1927.

Numbers, never great, had been dwindling: six registered in 1925, nine in 1926 and two in 1927. Most were registered by the Misses Window Harrison, whose 'Of Weald' herd later became famous for British Saanens. Certainly the two or three pints given by the old English milking goats must have seemed meagre indeed when compared with the yields of 1½–2 gal given by the improved goats.

But was the total abandonment of our indigenous British goat wise? She had excellent conformation and stamina, was able to pick a living and yield a quart of milk on scrubland and did not, as far as the writer has been able to discover, suffer from hormone imbalances and cystic ovaries, or other ills which make breeding chancy or impossible. She did not give birth to inter-sexed kids, a grave failing with the early imports of Toggenburgs, in whose country of origin hornlessness was considered a blessing—in fact, the only state! The officially recorded highest English yield seems to be that of *Elffye of Weald,* in 1929–30, with 1539 lb 8 oz.

APPEARANCE

The Old English which the author herself met with in the Cheviots were of small stocky type, with long sharp horns which swept back and outwards at the ends. Their heads were neat and wide between the eyes, with faces dished and mouths small. Their coats were most often blue roan, or fawn roan, sometimes with white patches, and in texture as well as in colour closely resembled the coats of the Scottish Bearded Collie, having a long straggly overcoat and a fine woolly undercoat. There were never any Swiss markings.

The ears were pricked and not wide; there was usually a beard of some length but no tassels. The legs were short and well boned, with fairly straight pasterns and neat hooves.

One of these goats which belonged to a smallholder in Sussex before the Second World War, gave birth to twin kids, her first, at the age of sixteen.

THE IRISH TYPE

At one time, Irish goats were coming into the West Country by the truck load, presumably because of war-time

Figure 6—Feral goat from the College Valley, Cheviot Hills, Northumberland

demand exceeding the local supply. These goats were of a quite different type to the Old English described above. They had short chestnut hair, were lighter in build, and gave more milk. Their horns were erect until the last few inches, when they turned straight back in a hook. They all, without exception, had the black eel stripe and black markings on legs, of the Highland, and Norwegian ponies.

THE SCOTTISH TYPE

Although earlier, there must have been some coloured Feral goats in the Highlands, the writer saw only white wild goats in the hills round Ardverikie and was told by Sir John Ramsden's Factor that Lady Ramsden had lent the Crofters a Saanen male which, of course, being dominant, would account for the lack of coloured goats in this district between the wars. There are probably coloured wild goats in other parts of the Highlands.

SURVIVAL

At the present time, efforts are being made to revive a small breeding nucleus of Old English goats in Lancashire; and the Rare Breeds Survival Trust has also obtained 29 of those recently culled from a herd, said to number 300, roaming the Cheviot hills and doing considerable damage. A British Goat Society representative contacted the Northumbrian Council and found that the National Trust were intending to select 23 females and three males for re-release in the hills.

Out of 114 Feral goats rounded up, 50 were slaughtered, but 64 were found new homes in National Parks and in private herds. It is good to know that the Northumbrian Dairy Goat Society is also taking an active interest. If those true to Old English type are selected and kept free of Swiss and of Nubian blood, it may well be that all is not lost, and an important animal has been saved from indiscriminate and total crossing with other breeds.

CHAPTER 8

THE GOLDEN GUERNSEY BREED

ORIGINS

Islands are always romantic, and the flora and fauna often exceedingly interesting, for it is not unusual for some species to have evolved which are not common to other places. Such is the Golden Guernsey goat, for no one really knows exactly how she got her colour, which is almost the same as the Guernsey cow.

Within the Author's goat-keeping lifetime, the lady who was primarily responsible for creating interest in the breed, Miss Milbourne, of L'Ancresse, Guernsey, began to search for animals of this colour amongst the British breeders and approached the Author for advice on the purchase of a stud goat for use on the small island goats, from which she hoped in time to achieve a line of pure breeding Golden Guernseys with stronger frames and higher production.

The venture proved fraught with difficulties, not the least being ill health, lack of permanent staff and restriction of the Common of L'Ancresse as a grazing ground.

Nevertheless, with the help of new and enthusiastic breeders who, by now, had formed the Guernsey Goat Society, and sponsored by the States of Guernsey Committee for Agriculture, who took up the cause of their National goat (after seeing a herd and finding the breed potential worthy of their support), breeding plans went ahead, and the interest of the British Goat Society and of buyers wishing to start herds in England, was at last aroused.

A Trust Fund was set up, and the British Goat Society set about bringing the breed into line with other evolved breeds, by opening a Register in 1975, for the resultant progeny of outcrossed Golden Guernsey females with Saanen or British Saanen sires.

The stock resulting from these outcrosses are mated back to Golden Guernsey or English Guernsey sires for three generations, and thus gain access to the English Guernsey Section.

Much voluntary work by members of the Guernsey Goat Society went into the Island's own Herd Book for pure-breds, and there has been good liaison between the two Societies. Shows, always keenly attended in the Island, have gone ahead well, and the progress made in the interests of the breed since 1922, when the Island Herd Book started, is excellent.

There remains, however, much work to be done on the animals themselves, and it may be some years before we see a supreme inter-breed champion Golden Guernsey.

APPEARANCE

The Golden Guernsey is a small compact goat, weighing about 120–130 lb females and 190–200 lb males. Her true golden colour must be skin deep, not just hair deep. She is the only breed to have erect ears with a slight curl at the tip, though this characteristic is not always found.

Her head is somewhat longer than that of the Toggenburg, but is dished, in the same manner, and whereas the Toggenburg almost always has tassels the Golden Guernsey does not.

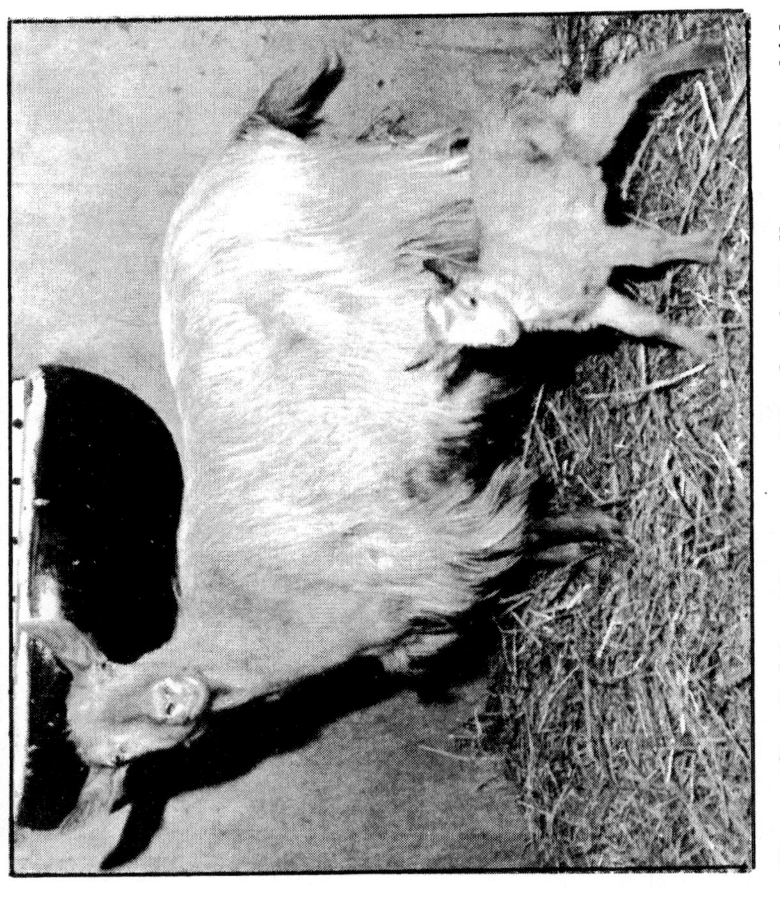

Figure 7—Golden Guernsey Westbank Holly and her kid.

Best in Show, *Rogate Club Show*

(Owned by Mrs Elspeth Riley-Smith)

The coat is usually short, often with a longer fringe down the spine and on the quarters. The legs are fairly light-boned, but straight and strong, with good pasterns and feet.

The whole appearance should be of an animal well able to produce a reasonable amount of milk on herbage, and like the Old English Goats this reasonable amount should be high in butterfats.

FEEDING

Winters are not normally severe in the Channel Islands and concentrates, which have to be imported, are costly so this little goat should not be improved as to milk yield to the point where she belies her situation, which is that of a grazing and browsing animal not a heavy, stall-fed creature.

NUMBERS

A census was taken in 1974 and the Rare Breeds Survival Trust believe that there are now some two hundred pure-bred Golden Guernseys. R155 *Novington Dorcas QX 1*, has recently given 3434 lb (1554 kg) at 4·61 butterfats. She has a show yield of 13·9 lb (6·3 kg) at 3·99 and 4·73 BFs.

BREED STANDARD

Head—Ears erect, slight curl at tip. Horned, hornless or disbudded. Facial line dished or straight.

Neck—Slender, without tassels.

Coat—Short, with or without long hair along spine and on quarters.

Colour—Golden, with or without white splashes. White blaize on head allowed, but no Swiss markings on face or legs.

Skin—Golden.

The following points differing from the ideal are found and are permissible: **Neck**—tassels; **Coat**—long; **Colour**—shades of gold, ranging from pale honey cream to deep gold; **Size**—larger than the ideal, but not coarse.

CHAPTER 9

PYGMY GOATS & MISCELLANEOUS BREEDS

PYGMY GOATS

This breed of diminutive goats is now much more in favour than the large, standard size breeds for many reasons, listed below. These goats originate in Africa, the most popular being the West African type, but there are other types.

Advantages

1. Quite small (around 37 - 45 cm tall) yet they are quite natural in their proportions.

2. Need less room for housing and grazing so a pair can be kept in a garden. Housing can be in a small loose box or even a garden shed.

3. Easy to handle and likely to be safer than some of the larger goats.

4. Liked by children and can become family pets. However, very young children should not be left with any goat – accidents can happen when an older person is not supervising.

5. Provide a good yield of milk which can be used by the household, whereas with the larger goats the production is so high that difficulties may be found in selling the produce unless properly organized on commercial lines.

6. They can be disbudded to prevent horns growing, but this should be done by a veterinary surgeon and this will usually be done under anaesthetic. It must be done before the buds have grown into horns.

7. Easy to keep and breed quite readily. They are quite hardy and, as a result, are relatively free from health problems.

Pygmy Goat & Kid

THE BAGOT GOATS

A special breed of goat which was originally owned by the Bagot family at Blithfield Park, but are now to be found in Wildlife Parks. They are black and white. Commercially they are of little significance, but are interesting because they are a very old breed having been in England over 600 years.

FOREIGN GOATS

There are goats to be found in many lands, many serving the community in providing milk or meat. They also provide skins for bottles and manure for cultivation. There are also many wild goats which inhabit areas, often hilly, and not really suitable for cultivation.

Goats are utilized in France, Spain, Denmark, Italy, Norway and America. France is noted for the fine cheeses made from the milk provided by large commercial herds.

The Cashmere goat is also much prized because of the fine silky wool that is used for beautiful garments.

THE ANGORA GOAT

Whilst the knowledge that the mohair obtained from the Angora Goat just as the wool obtained from the Merino sheep, was apparently badly affected by our climatic conditions in the British Isles, it is only quite recently that suspicion aroused in the minds of would-be owners of the Angora have caused them to import again these animals which, for so long have been written off as un-economic here.

The suspicion is that it was the goat husbandry itself which may have been at fault, not the actual growth of mohair. Time will tell, as there are now sufficient numbers of these animals in different surroundings, to make a reasonable assessment of their chances, within the next few years. Whether they will become a viable commercial breed remains to be seen, but they must be given a fair chance now, for their earlier importation, back in Holmed Peglar's day, cannot rule out the possibility that it was the goats themselves which, by our present day and more knowledgable standards, were badly kept.

The writer has had the privilege recently of seeing Mrs Rosenburg's herd at home on her Rare Breeds farm, and they are, as one can see from the one pictured overleaf, outstandingly beautiful, with extremely good and sound conformation. Perhaps this latter may be in some measure due to the fact that no-one has tried to up-grade them for milk production, at the expense of sound conformation, a short-sighted policy from which our dairy goat breeders are not, unfortunately, free.

The yield of hair, which is shorn twice annually, is said to be about 10 lbs per goat. They are also useful producers of kid meat, as these latter put on flesh rapidly, a point often missed by our dairy goat breeders, since few dairy breeds can honestly be considered dual purpose.

Angora Goat
(owned by Mrs M. Rosenburg)

CHAPTER 10

THE MALE GOAT

IMPORTANT ASSET

The male goat is at one time the most important and the most secluded member of the herd. The former because he will be responsible for all the offspring during his reign, whereas each female will only be responsible for her own, so that his breeding and his performance must be of supreme importance to the serious breeder, and it will be essential when buying an adult male or when using one at stud elsewhere, to study the conformation, the milk yields, the length of lactation and the butterfat content of his daughters, as well as the breed type.

There is considerable confusion as to the meaning of the words 'Proven' and 'Proved', both used indiscriminately in advertisements. 'Proven' means that a male has sired daughters which are better than their dams in yield, type and perhaps butterfats too. 'Proved' simply means that a male is capable of getting stock.

CHOOSING A MALE

It is so easy to breed down, rather than up, in any form of livestock, and it is so absorbingly interesting to try to improve within one's chosen sphere of livestock breeding, that few will not take the trouble to seek out the male goat best suited to their foundation stock.

Some misunderstandings also exist with the terms 'line-breeding' and 'in-breeding', and the simplest way to explain the difference to the novice is by way of pedigrees which are examples of each. It is not sufficient to use the nearest registered male of the same breed, nor is it enough to choose one of the same breed which is an out-cross with regards blood lines, compared with the foundation goat which is to be served. This may result in a lucky fluke, but it will in no way fix those characteristics which are desirable in the goat. To do this the pedigree must be studied, to find out which are the most successful ancestors, by way of Herd Book milk records, show reports, etc.

YOUNG MALES

It frequently happens that because of difficulty in finding, obtaining or even in managing an adult male (who may well be a tough customer for an elderly or small lady to manage), a male kid must be used or purchased. Here the distinguishing signs awarded to males whose dams have qualified for an R by virtue of Milk-Recording; or a Star or QX, for points gained in a recognised milking competition at a show, are of great help in selecting a potential sire of good milkers. Unfortunately, the sure sign of a good animal for producing good milkers amongst his progeny, the Sire of Merit (SM) award can obviously only occur in adults, and sometimes posthumously.

APPEARANCE OF THE MALE GOAT

A diversity of opinion exists as to whether the best male goats should look male or female, and 'milky'. The writer prefers those which appear entirely masculine, with good stout limbs, one 'at each corner' like a well-built bull.

This does not mean a coarse animal, but rather one undeniably the leader of the herd, and he should have elegant fine shoulders, a level back and an intelligent and bright eye, in a masculine but attractive head of good breed type.

A disbudded male kid is a safer rearing proposition than one which is naturally hornless, because the hornless condition in goats is sometimes linked with sexual abnormalities, and a hornless male kid is more likely to prove sterile than one which was naturally horned. However, males are very difficult to disbud absolutely 'clean', and some small 'slugs' or horny growths about half-an-inch long, with blunt ends, and which often knock off after a fight, are often left. These should not make any difference to the sale or show value of the kid. Try to avoid a sharp point which could be a dangerous weapon, and for this reason it is usually best to disbud male kids at two days old, instead of the usual four days for females.

It is said that a male with well hung, well developed and level testicles, is one which is more likely to produce udders of good shape in his daughters.

LIMITATION OF SERVICES

It was once supposed that a male kid would be ruined if he were allowed to give more than about six services during his first season, but it is not the numbers of times he serves during this time but the interval between services which is important. A six-months old male should certainly not serve more than one female in a day, but he is quite capable of serving some thirty goats during his first winter if he is fit, well exercised and well fed, and if these services are well spaced out during this, his first season.

SMELL AND SECLUSION

The male goat is endowed by nature with a distinctive and, most people think, an abominable smell, designed to attract his females from afar. This is most noticeable during the rutting, or breeding, season, which extends from late August to mid-March. Some breeders have their male kids 'descented' at the same time as they are disbudded; an operation consisting of burning down the scent glands situated behind the site of the horn buds. It is not clear as yet whether this is entirely successful, since part of the objectionable male goat smell undoubedly comes from his unpleasant habit of spraying himself with urine, which is also extremely obnoxious.

Anyone thinking of allowing children to keep a pretty little male kid as a pet, without having it castrated, will one day regret the idea, for a more *unsuitable* pet than a grown male goat would be hard to come by.

Whilst it is a good plan to have the male's house and exercising yard within sight of the female herd for his interest and contentment, it is a bad plan to have him within smelling distance of visitors, for they immediately get the idea firmly fixed in their minds that goats smell horrible and this, of course, is very far from the truth, as the female goat smells of hay and a variety of leaves and grass, all pleasant and acceptable to the most finickity nose.

The attendant's clothing must be protected when he goes to feed or take out the male for a service and plastic gloves prevent the hands persistently smelling. A bull stick does help very much with the handling, making it impossible for the male to reach the attendant and easy for release without touching the animal by reason of the remote control clip;

this of course is also clipped on to his collar, or head collar, when he is taken out and before his door is opened.

AGE OF SECLUSION

A male kid is capable of effective service between three months and five months of age and must, thereafter, be housed separately or there will almost certainly be stolen matings, even between the male kid and his erstwhile companions of the same age.

Another point of great importance, particularly where milk is sold, is the tainting of the milk with the male goat smell. Avoiding this problem means following strict procedure of putting on a plastic 'mac' or overall before entering the male house, and removing it on leaving. It may even be necessary to abandon the milk of a goat which has just been served (for animal feeding) because the male has rubbed his head on her flanks and the smell has spoilt the first milking afterwards.

One sees far too many cow-hocked male goats, and although this is probably environmental and is brought about by lack of exercise, it remains a fault to be avoided.

REARING A MALE

The young male kid of a few weeks old when purchased should at first have not less than 4 pints of milk daily, in four feeds, and although the milk feeds may be reduced gradually, so that he is getting two feeds or a total of about 3 pints daily at four months, he will need fresh greenstuff daily to capacity, to make up for his having to be stall-fed. He will have been introduced to small concentrate feeds from the age of three weeks, and by six months should be taking about 1 lb daily. Variety is extremely important, three small feeds being better than two larger feeds, for during the rut his mind will be on his work and he may need tempting. If one bottle of milk can be continued throughout his first winter it will greatly benefit him, helping to keep him growing and in good condition.

Hay is of the utmost importance and, whichever type—meadow, clover or seeds mixture—it must be the very best obtainable and must not be stinted. Swedes and carrots, and soaked sugar beet pulp are all fed to male goats, but not mangolds, as these can cause silting up of the urinary tract,

<table>
<tr><td rowspan="8">Bassett</td><td rowspan="4">Tzaraf</td><td rowspan="2">Dandy</td><td>* Shadoof</td></tr>
<tr><td>Alisa</td></tr>
<tr><td rowspan="2">Almond</td><td>Luther</td></tr>
<tr><td>Ch: Allsorts</td></tr>
<tr><td rowspan="4">Lickrish Br: Ch:</td><td rowspan="2">* Shadoof</td><td>Shifari</td></tr>
<tr><td>Georgette</td></tr>
<tr><td rowspan="2">Peire</td><td>Prefect</td></tr>
<tr><td>Pennywise</td></tr>
<tr><td rowspan="8">Ch: Aniseed</td><td rowspan="4">Luther</td><td rowspan="2">Subadar</td><td>Harkaway</td></tr>
<tr><td>Pandora</td></tr>
<tr><td rowspan="2">Wrishka</td><td>Sancho</td></tr>
<tr><td>Sasha</td></tr>
<tr><td rowspan="4">Ch: Allsorts</td><td rowspan="2">* Shadoof</td><td>Shifari</td></tr>
<tr><td>Georgette</td></tr>
<tr><td rowspan="2">Peire</td><td>Prefect</td></tr>
<tr><td>Pennywise</td></tr>
</table>

Diagram 10-1—In-breeding illustrated—bringing in the successful sire *Shadoof* 3 times*

Ch: Ligonier	Ch: Barry	Prune	Landings
			Nymph
		Spark	Balaize
			Spode
	Ch: Lily	Bartsia	Sovereign
			Pattern
		Shoes	Balaize
			Spode
Ch: Lucia	Ch: Controller	Fox	Marksman
			Perpetual
		Mist	Story
			Skylark
	Ch: Lady	Barry	Prune
			Spark
		Lily	Bartsia
			Shoes

Diagram 10-2—Successful line-breeding. *Ligonier* sired 18 champions including *Barndance.* (This is the latter's pedigree.)

which can prove fatal. For this reason, in a hard water area soft or rain water, is better given to male goats than standard tap water.

BEHAVIOUR

Never allow children to play with a male kid, the playful hopping and butting can turn into something far removed from play in the adult, which may weigh 2–3 cwt and be capable of knocking a man down. Firm but kind discipline at all times, lead training and a routine of going out to exercise is desirable.

ARTIFICIAL INSEMINATION OF GOATS

In other, larger, countries, where transporting the female many miles to the desired male may be a problem, artificial insemination may be used fairly extensively; in the British Isles, however, where the distances are hardly ever farther than 150 miles, it is seldom considered. Success rates vary from about 60 to 80%.

It is quite usual for British breeders to pass round a favoured male goat, allowing him to stand at stud in different areas, and this practice has much to recommend it. The stud goat owner will normally issue a certificate of service to the female's owner, which must be kept carefully for the registration of the resultant kids later. If the goat is on loan, the borrower should sign in the space provided as 'Authorised Agent', and the owner should have previously informed the Secretary of the British Goat Society of the transaction.

RETURN SERVICE

It is usual for the stud goat owner or agent to allow a return service, free of further fee, three weeks after the first service if the female returns. He is not bound to allow further services, however, if the female again returns.

STERILITY IN MALES

As has already been said, male kids may be found, on test, to be sterile, and it is most frequently those which are naturally hornless, but it could be the result of an accident.

An old male will often become sterile and very little can be done to improve this situation, unless it is simply a matter of poor management.

Like the females, a male goat should have routine worm doses and his feet, which are all-important, should be trimmed regularly about every four or five weeks.

The paragraphs below are an example of successful in-breeding and line-breeding:

> "The Chillingham Herd of White Park cattle have been in-bred for seven hundred years, with no apparent ill effect, except for some loss of size.
>
> Most problems occur early in the in-breeding programme, when undesirable recessives may come to the surface.
>
> If this phase can be negotiated successfully and the bad points bred out, the resulting animals will be of superior genetic merit, and prepotent for their quality."

(G. L. H. Alderson, in *The Ark*, the Journal of the Rare Breeds Survival Trust, reproduced by kind permission.)

One must always remember, however, that in breeding domesticated animals there is no question of survival of the fittest—as with the Chillingham herd, for example, or the Feral goats. There the inferior or weakly males would go down before the more robust and virile animals, and this makes it doubly important that no lack of stamina or faulty conformation should exist in the males chosen, especially for an in-breeding, or a line-breeding programme. **Diagrams 10-1 and 10-2 illustrate in-breeding and line-breeding.**

CHAPTER 11

PURCHASING STOCK

FIRST STEPS

Often the householder who seeks to keep a couple of reasonably good and realistically priced milking goats later becomes an enthusiastic exhibitor. It is suggested, therefore, that the first approach should be to make enquiries through the local affiliated goat club and the British Goat Society's monthly journal.

REGISTERED STOCK

It is not always true that you get what you pay for, but it is certainly advisable to buy the best that you can afford, because the registered milker with known pedigree for production, will cost the same to maintain as the unregistered scrub goat bought through the local newspaper and costing a few pounds. The difference will not only be the pleasure of looking at, and working with, a beautiful animal, but the improved yield of milk, and more important still, the longer length of lactation of the one which has been bred for generations with this in view.

CONFORMATION

Assuming that the chapters on the breeds have been read and a decision arrived at as to the most suitable breed, and that the method of management is decided, the question of housing settled, and the availability of hay, bedding, concentrates and greenstuff ascertained, a clear idea should be gained of the conformation of a good milking animal which is common to all breeds.

If the pedigree is right and the conformation good, there remains the question of soundness, health and age.

TELLING THE AGE BY THE TEETH

A goat has a hard pad in the front of the upper jaw, with six molars on each side of the upper and lower jaws and eight incisors below. The milk teeth are replaced by permanent ones by degrees, as shown in the diagram of the ages. At five years old a goat has a 'full mouth'. After this age, the state of the teeth must be the guide as to age. They become worn, and eventually some are lost. The goat is then too old to be a suitable purchase.

AGE TO BUY

The ideal to aim at would be a reared kid, with an in-kid goatling, because the new owner will have time to become accustomed to the animals and have gained a fair idea of their routine, before the goatling has her kid and becomes a milker. The kid will act as a companion, and will be ready to mate the following autumn. Goats are by nature herd animals and are not happy alone.

If both animals can be bought from the same herd so much the better, for the male goat eventually chosen will probably suit both and they will certainly settle down together more quickly.

Registration and Transfer forms

A goat must be registered by its breeder (unlike a puppy) and consequently it is essential to obtain the registration card and transfer card from the seller, or the animal will be officially non-pedigree, however well-bred.

Export Certificate

Foreign buyers who are unable to inspect stock for themselves, may enlist the services of a licensed British Goat Society judge, who will, for a reasonable fee, inspect intending purchases and report on type, soundness, and correct papers. The goat must be earmarked for obvious reasons. There are also reliable breeders who will handle all arrangements with the Ministry of Agriculture, and shippers.

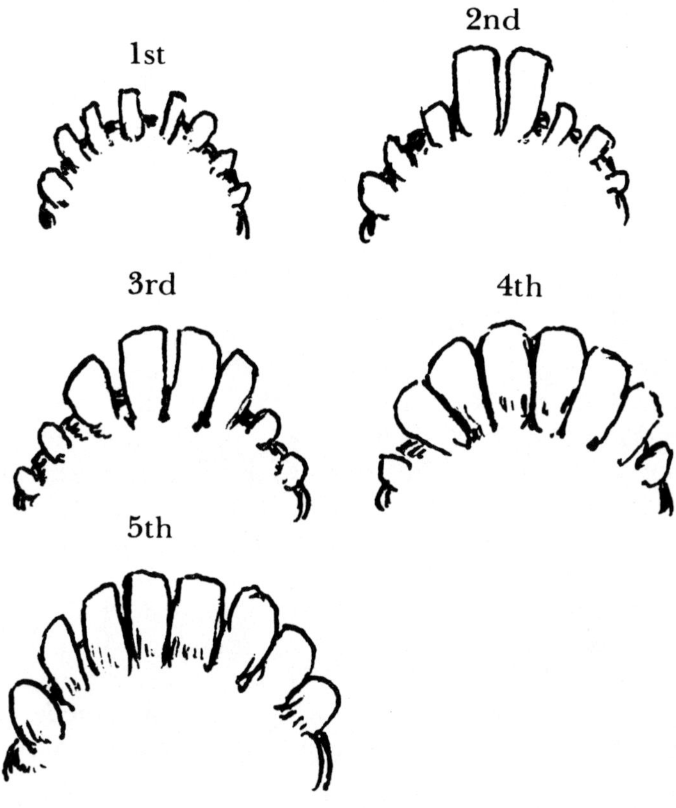

Diagram 11-1—Age of a Goat By Its Teeth (Years 1–5)

DISTINGUISHING SIGNS

These are awarded by the British Goat Society for milk production in recognised competitions at shows and for officially recorded yields over the goat's first 365 days of lactation and can be studied on her pedigree—they are listed and explained in Chapter 16.

Thus a goat with some 'decorations', which has been mated with a male whose dam has also qualifications, is likely to produce worthwhile stock. A study of the breed *standards*, together with the average and the highest yields, will give the buyer some idea of perfection, but this is seldom for sale and, in fact, rarely achieved by any breeder of livestock.

PRICE

Inflation operating whilst the writer is endeavouring to give the prospective goatkeeper some guidance on price makes this well-nigh impossible. Suffice to say that, in 1970, one could buy a fair specimen entered in the Herd Book and giving around a gallon of milk daily, for about £30, or a well-bred registered kid of some six months old, for about £15. Now the price is much higher.

To bring the milker to profit will now cost something in the region of £80 to £100, owing to the tremendous rise in the cost of feeding stuffs. It follows that the show specimen capable of winning at the bigger shows and a potential breeder of stock which will make good prices abroad, will cost upwards of £180.

FAULTS WHICH MAY BE CONSIDERED LESS DETRIMENTAL

It would be better to buy a well-bred, well-reared animal with a fault which might debar her from winning at a large show, having paid due regard to her forbears and ascertained that this fault is not fixed in her family. This goat might, if mated to the right male, produce for the novice something exceptionally good.

Slight sandy shadings on the white parts of the Toggenburg, British Toggenburg and British Alpine, differ from the ideal, but are not disastrously impossible to eradicate in future generations.

A slightly wry tail in an otherwise good Anglo-Nubian with a good mouth, could be forgiven, as a foundation goat. So also might a shade of Swiss markings on the face.

Very small or very large teats, neither of them practical, might be suffered with the hope that an otherwise good animal could, if served by a male whose female forbears all had excellent teats of the right size and shape, give birth to progeny with exactly the right size of teat for ease of milking, attached to udders of spherical perfection.

Perfection all-round is well-nigh impossible to achieve, still less to purchase, but the Author prefers a well constructed animal with a moderate amount of milk to a very heavy milker whose frame is already beginning to show signs of sagging under the strain.

BRITISH SECTION OF THE HERD BOOK

Because those goats entered in the British Section of the Herd Book may look exactly like animals entered in their respective breed sections, and are in fact frequently advertised as 'Registered', this can make for confusion for the beginner who should always see the Registration Card before purchasing. At the top right hand corner of this card a breed registered goat, as opposed to a Herd Book registered goat, will have its breed initials before the number, thus 'AN2352. BT6547', and so on, whereas the British Goat will have only 'HB' before her number. She is in fact unique, because in no other form of livestock is an animal which is a crossbred, or one which is grading-up, given British Nationality!

It therefore follows that an animal registered in the British Section of the Herd Book is worth rather less than one in the Breed Sections, and it very seldom happens that an export enquirer will accept a 'British Goat'; nor are males thus registered very easy to sell.

If all else fails and the buyer finds himself unable to obtain exactly what he wants at the right price, he should look carefully at the pedigree of the British Goat and count up the generations of grading, with the table in the British Goat Society publications, to help to find out for himself how many times the goat's progeny and the resultant progeny must be put to a male registered in the appropriate Breed Section, before his home-bred goats eventually achieve their section. As already stated, the pure breeds, Toggenburg and Saanen, have a closed Herd Book, into which no grades can ever be admitted.

The skill of certain breeders in introducing British blood into breeds which have no pure stock to which they may return (as, for example, Toggenburg and British Toggenburg, Saanen and British Saanen) has made improvements in certain lines, but it is not recommended that the novice, with no knowledge of just what has gone before, should attempt this kind of outcrossing.

THE HEALTH OF THE PROSPECTIVE PURCHASE

A healthy goat has bright eyes, a soft flexible skin with no harsh hair (no matter whether this be long or short), sound teeth and good, well-shaped feet which have been trimmed regularly (they should be neat and compact, not twisted or spreading). If a milker, the udder should be soft and free from lumps or skin troubles.

If the goat is heavily infested with internal parasites and is, in consequence, anaemic, the skin under the tail and the eye membranes will show as too white.

If it is possible to see the goat milked at two consecutive milkings a fair idea of her yield may be gained, also whether she is quiet and easy to milk. Her milk should be tasted to test it for off-flavours, some of which are basic and incurable, whilst others may simply be caused by unsterile utensils, dirty udders or faulty timing of feeds.

A bad disbudding leaving 'slugs' growing, though unsightly, is not a fault which would debar the animal from exhibition, but a disbudded goat is preferable to one which is naturally hornless, on account of the fact that hornlessness is linked with the inter-sexed, so-called hermaphrodite, state (see 'Breeding').

SELECT CAREFULLY

One of the worst pitfalls for the novice buyer, is feeling sorry for some poor creature which has been spending its life miserably tethered out in the rain and at the mercy of swarms of flies, in hot weather, with no shelter and little food. One cannot make a silk purse out of a sow's ear, nor can one by magic and however much kindness and good subsequent treatment turn a 'scrub' goat into a profitable

THE POINTS OF A GOAT

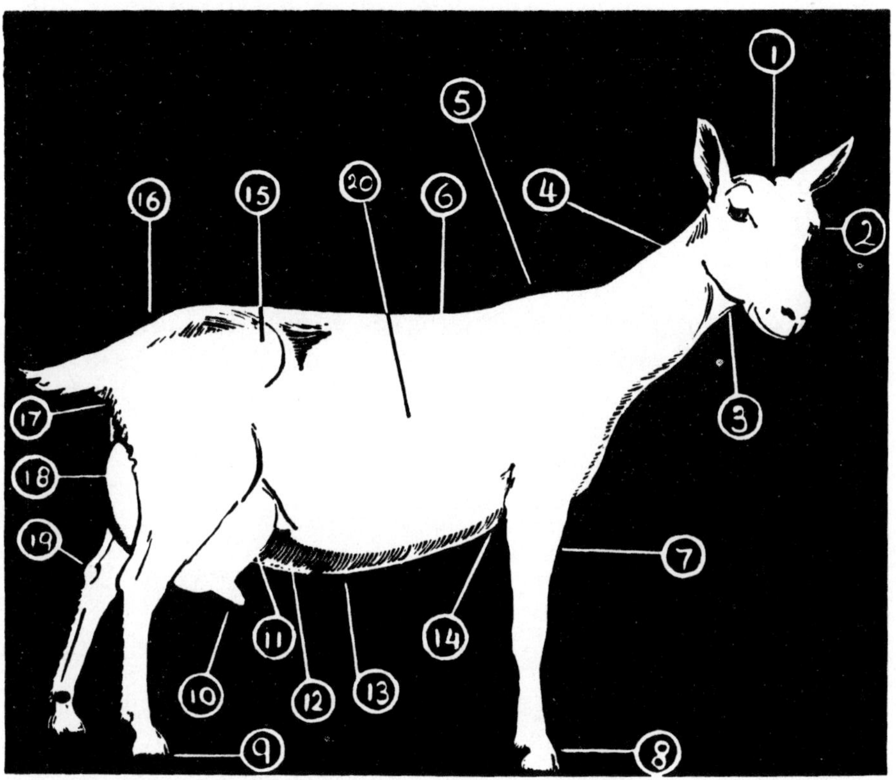

CONFORMATION FOR PRODUCTION

1. Head, pleasing, placid and intelligent.
2. Eye, bright and gentle.
3. Throat, clean and fine.
4. Neck, long, "milky" and not coarse.
5. Shoulders, clean and neat.
6. Back line, long and level.
7. Fore-legs, straight and sound, not too close.
8. Pasterns, fairly straight.
9. Feet, sound and neat.
10. Teats, tapering and pointing slightly forward.
11. Udder, spherical, firmly attached, silky, not fleshy.
12. Milk veins, prominent and tortuous.
13. Barrel, showing capacity.
14. Body, deep, allowing heart-room.
15. Pelvic development, broad and wide.
16. Rump, sloping gradually, not dropping suddenly.
17. Escutcheon, wide and reaching high.
18. Rear udder, developed, balanced.
19. Hocks, wide apart and straight.
20. Ribs, deep and well sprung.

proposition. Naturally, some improvement can be made by worming, gradually introducing good food and management. Grading-up from a scrub takes years. For several generations the progeny will be valueless, so turn resolutely aside from that poor old thing with the pretty kids running at her side, then keep firmly to the plan to buy only registered stock which will increase and not decrease in value, and which will give a good start if the hobby of exhibiting begins to appeal after one or two local shows have been attended.

Always remember, when looking for foundation stock, that if you are offered a fairly ordinary but sound and typical specimen which originates from an exceedingly good line, within its breed, it is a far better proposition than one in the first flight at the shows that is the result of a lucky chance mating from quite an ordinary family. As a breeding propositon, if properly mated with attention to the most prepotent male for all that is best, there will be a good likelihood of producing something really excellent.

CHAPTER 12

FEEDING

ESSENTIALS

A goat's food may be divided into two rations—the maintenance ration and the production ration. For young stock, provision must be made for strong growth as well as maintenance, although there is no milk production to cater for, as yet.

MAINTENANCE

To keep alive and in good condition, each adult goat will require roughage and concentrates. The fresh, young spring grass of a well-managed pasture will be eaten with relish and will provide for maintenance and about 2 pints of milk. Once the summer is at its height, the protein in the grass declines and must be made up if the milk supply is not to diminish beyond recall.

HAY IS THE MOST IMPORTANT FEED

Since goats are usually housed at night, they are then fed hay, even in the summer, and about 4 lb per head should be reckoned upon from May until August, when up to 7 lb might be eaten by a large goat giving 1½ gal daily.

Types of Hay

It is useless to economise when buying—always get the best hay of its kind obtainable. Bad hay will be wasted. Mouldy hay is downright dangerous.

Exhibitors usually feed all-clover and/or lucerned hay, with a little herby old-pasture meadow hay as a change. It is unwise to feed only clover hay ('stover' as it is called in the Eastern counties); when too much green leguminous freshly-cut food is given this may upset the balance.

Various plants may be made into hay: nettles, docks and comfrey are all used at times, but will only dry out in really hot sun.

Lucerne, clover and sainfoin—the legumes—should really be tripoded (*see* chapter on 'Crops to Grow').

Preventing Bloat

It is important to feed hay before turning goats out on to a new pasture, or before folding on kale, lucerne or other green crop.

Hay is fed in racks, or nets, but the latter are not safe for use where there are very young kids, which jump up and easily become entangled, and have been known to hang themselves. The type of hay rack illustrated in the chapter on housing will save much waste. For show-going a small portable rack is used. Hay thrown down upon the floor will be trodden on and then refused. All utensils must be kept clean.

Three Times Daily

In wet weather, or in winter, hay racks are replenished three times daily, and in summer in the early morning and last thing at night.

Although the bigger herds will require summer crops to be grown for feeding on wet days when the goats cannot go out to graze, and at shows, the household couple can be fed branches of elm, ash, willow, hazel, sweet chestnut, ivy, holly, hogweed, gorse, brambles and raspberry and other fruit prunings, which may usually be found in adequate supply seasonally, in the garden. There will also be pea and bean haulm and brassica stalks, and outside leaves, for use in the goat house.

PRODUCTION RATION

Thus the maintenance of the goats is supplied, and the 'factory' must be given the raw materials for making the milk, according to the yield or potential yield, of each goat. It is usually accepted that a goat, or a cow, requires 3½–4 lb of a balanced dairy concentrate ration for each gallon of milk given. Young spring grass may account for 2 pints, and this is also true of the legumes, kales etc., but a goat's ability to give a good quantity on greenstuffs will depend largely upon the individual—very small animals not having the capacity for the large intake of bulk required.

Commercial Mixtures and Nuts

In some herds, dairy nuts are popular; in others, the goats soon tire of them. Most breeders have their own mixture of cereals and protein cakes made up by their miller, although some are able to feed one firm's coarse dairy ration indefinitely, without monotony.

A dredge corn seed mixture may be obtained and sown, if there is sufficient land and enough stock. The resultant concentrate mixture, when milled, will be balanced, and will require only an adequate mineral additive. This home-produced grain ration also has the advantage of not including fish meal, greatly objected to by goats. Horse stud nuts are sometimes preferred to dairy nuts, and since intended for lactating mares, are reasonably suitable, although not the equivalent of dairy nuts. Horse and pony nuts contain a useful amount of grass meal, which may not be liked by the goats in the dusty state; lucerne nuts may also be bought. The latter are useful at times when there is no grazing or for stall-fed goats.

As will have been noticed, "one man's meat may be another's poison", so it is largely a matter of personal preferences, always providing that the concentrate ration used for milk production is properly balanced.

Cheaper By-products

Part of the production ration may be fed in the form of sugarbeet pulp, a by-product from the sugar factories. This was once 8s per 1¼ cwt, but is still less expensive than cereals. The beet pulp, or beet pulp in nut form, should be soaked in boiling water until it has swelled out. When cool enough to handle, it is dried off with a little broad bran, if liked, and fed at the rate of 6 oz per goat *weighed dry*. It is important to soak beet pulp in a plastic container, not a galvanised one. Rubber gloves for handling the soaked pulp are a good idea, as it tends to chap the skin. Beet pulp is mollassed after processing and the excess liquid is usually appreciated as a drink—it is a good preventative of Acetonaemia.

A small amount of dry beet pulp may be added to the coarse dairy ration, but watch must be kept on goats inclined to bloat and in-kid animals, which may not be able to digest the feed on account of its great swelling when wet.

So, with concentrates, green feeds and hay, there remains only the root feeds to mention. These, in the fresh state, may consist of (in order of priority) carrots, fodder-beet, swedes, mangolds, parsnips and turnips. They are fed cut into slices or cubes, and sometimes dusted with bran. A simple home-made root cutter is illustrated.

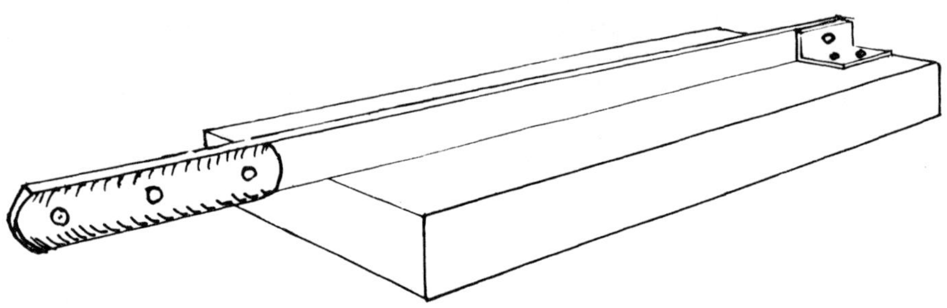

Diagram 12-1—Root Cutter

Mangolds are clamped until after Christmas, and the leaves of all roots are best ploughed back into the land. If 'chat', or small potatoes, are to be used, they should be boiled and the water thrown away. Never feed green potatoes.

A perfect sample of silage will be eaten willingly by the goats, but this is very difficult to make well in small quantities. The writer found that less than 12 tons was wasteful round the edges, at the top and the bottom of the silo, there being the same margin of waste in 10 or 20 tons.

Home-made concentrate mixtures will require some form of mineral supplement—here it is a matter of trial and error, for some herds will take to one brand, whilst others will take only another. *Equivite,* which is a mineral made for horses, seems very popular and can safely be added to each goat's feed, once daily, at the rate of one level desertspoonful. It is said that it is insufficient in some respects, but it is really a matter of finding out what is in short supply in the particular district, some areas being low in calcium whilst others are low in cobalt or iodine. Within a few yards there may be such a difference that it is worth making enquiries. Find out if the dairy herds in the district suffer from any deficiency diseases. The soil may be tested.

The following mixtures of cereals and protein cakes or nuts are all balanced for milk production, and may be fed at approximately the rate of 3½–4 lb per gallon of milk. If two or three of these mixtures are kept in good dry storage bins, perhaps in ½ cwt lots, so that they are not kept so long that they go stale, the changes may be rung and the animal's individual fancy catered for—a practice which will be handsomely repaid in the milk bucket. It may even be that the best milker prefers to drink out of a green bucket, whilst her sister prefers the white one. By pandering to these small preferences more water will be drunk and, in consequence, more milk given.

WARM WATER, AND SOME VARIATIONS

All goats drink more water if it is heated to just above blood heat. Some like an occasional teaspoonful of salt in the bucket, too; others like oatmeal, and most like the sugarbeet soaking water. If water is not left with the goats, they must be offered a drink three or four times daily.

Time for Milking

Milking would take place during or after the morning and evening concentrate feed. Any food left over should be removed and the amount reduced, or the type changed, according to yield.

A mineral lick brick should be kept in a holder in each goat's pen.

When feeding kale, it is best to retain the root so that the plant may be hung up. This wastes less than cutting and feeding in nets or racks. This also applies to cattle cabbage (Flatpol).

A DAY'S RATIONS FOR THE GALLON MILKER*

In Summer

6–7 a.m.—1½ lb mixture No. 1; hay, fresh warm water.

About 9 a.m.—out to graze.

Midday—water if none available in field.

6 p.m.—return to house; fresh warm water; hay ready in racks; 1½ lb mixture No. 2.

8/9 p.m.—greenstuff or branches as available for over-night.

Note that in very hot weather when flies are troublesome, it is best to bring the goats in and feed branches, or other greens, during the heat of the day. Entirely stall-fed goats will need greens at midday. Very high yielding goats will also require soaked beet pulp.

In Winter

7–8 a.m.—2 lb mixture No. 1; fresh warm water; hay; greens as available.

Mid-day—soaked beet pulp *or* cut roots with broad bran; water; hay.

6–7 p.m.—2 lb mixture No. 2, or dairy, or stud nuts; water; hay.

9 p.m.—kale.

Introduce Different Foods Gradually.

It is virtually impossible to over-feed branches, the goat's natural diet, but when kale and roots come into use later in the year always start with small amounts and increase gradually; always feed hay first.

* The mixture numbers (Nos. 1 and 2) are given by way of illustration. *Variety* is the important theme. Possible mixtures are given at the end of the chapter.

Folding Goats on Fodder Crops

A plot of kale or fodder mixture may be grazed by moving the electric fence daily, allowing just as much as the goats will clear up readily, but there should always be access to old pasture, or to hay, to offset the lushness of the folded crop. Alternatively, the goats may be moved away after a short spell and returned for a time later in the day.

Winter grass, unless a modern ley specifically adapted for use at this time, cannot be regarded as part of either the maintenance or the production ration, but only as a change of scenery and a place for exercise. Heavy mud may necessitate stall-feeding the goats but a daily walk or exercise in a yard is extremely important, particularly for those which are in-kid.

CHANGING FROM WINTER FEEDING TO SPRING GRAZING

As already stated, any change must be gradual, and no less so because it is a natural one. The change from winter to spring grass would not, however, come all at once in nature and it will save digestive upsets and metabolic disturbances if the herdsman takes one of two alternative precautions.

First, if the goats are actually on the pasture when the spring comes, the first flush of grass will be gradual and their diet, therefore, gradually enriched. Again, if the goats are fed good hay immediately before turning out, and are then only left on the new pasture for about twenty minutes the first day, increasing the time allowed every day, they will take no harm.

There is so much variation, even within the breeds and in individual specimens, over the question of over-eating that no rules can be laid down. The herdsman should keep a watch on the state of each goat's figure and its droppings, thus avoiding trouble. It is particularly important to know the animals when taking them long distances to shows; some will hop in the van and leave readily at their destination none the worse, others will lie down and stay down too long, without cudding, so that they become blown. These goats should be taken any journey on a comparatively empty stomach; that is to say, feed no concentrates, but plenty of hay and branches. On arrival, they should at once receive hay, then water and branches, but no concentrates until they are seen to resume cudding, probably at milking time.

CONCENTRATES FOR KIDS

Whilst the young kids are still bottle fed and taking about 2–4 pints of milk daily (even if this is skim milk) they do not require the 16 or 17% protein ration which the milkers need, as they are getting sufficient from the milk. Once on dry feed only they require a balanced rearing ration and this may be in the form of calf pencils or calf ration (coarse).

During the milk-feeding stage, a little flaked maize and broad bran, with a pinch of mineral mixture added, is sufficient, this may be fed twice daily before the kids have their bottles.

Hay and Greens for Kids

Sweet-smelling, fine, herby meadow hay is ideal for young kids and they will begin to eat this at a week to ten days old, when they will soon be seen to chew the cud.

Their first greenstuff is usually branches, but if very early born they may learn to eat a little kale; however this must *never* be fed to any goat frosted, so, in very frosty weather, it is usually got in a day in advance and allowed to thaw out.

Bottle Feeds

Kids are usually removed from their mother's pen, whilst she is taken out for exercise, three or four days after she has kidded. The reason for this is that the kids thereafter regard the human attendant as their mother, are tame, friendly and easy to handle. On the dam's side, there is the decided advantage that her milk is taken completely, twice daily at regular hours, and she does not develop the annoying habit of holding it back for her kids. One can sometimes feel the goat stop letting down her milk and she will not respond to any bribery or coaxing until the kids are returned to her.

This has the effect of reducing her yield, for as the kids grow up she will gradually give them less and they will eat more solids, whereas, if she is milked and the kids bottle-fed after the third day, when they have had the all-important colostrum, or first milk, from her direct, both parties will do better.

The Value of Colostrum

Colostrum contains anti-bodies which give the young kids a certain amount of immunity to the various germs present in the soil and the goat house. If there is an excess amount of

Diagram 12-2—How To Carry A Kid

colostrum, it may be frozen for use in an emergency, such as for an orphan kid. Any not required may well be fed back to the goat, as it helps to prevent deficiency of calcium and inabiltity to absorb calcium, so preventing milk 'fever'.

New-born kids may be rather unsteady and not able to find their dam's teats at first, particularly if these are rather large and low. They soon learn if they are helped; if they are somewhat weak from a multiple or delayed birth they may be given a bottle feed to help them get on to their feet.

Warm Bottle with Wide Top

For this the Author uses a Maw's baby bottle with a wide top, into which one may milk direct. The teats which come with this bottle are soft and the correct size. It should be filled with hot water until immediately before the milk is drawn straight into it.

The kids are placed with their backs to the attendant's knees, or if weak, on the attendant's lap, facing away. They soon learn to suck, and after the third feed will come of their own accord.

Four Feeds Daily Until Three Months

Never force milk upon the kids, but let them have as much as they will take readily. At first this may be 3 oz or less, but by the time they are three months old they will probably be taking a full wine bottle, which is about 1½ pints or a little less, and at this stage cut their bottle feeds down from four to three daily, increasing their cereal ration gradually.

At Four Months Three Feeds and at Six Months One Bottle

At three or four months the kids may, if milk is short, be put on to a proprietary milk substitute—be careful to follow the maker's instructions, exactly, and always keep the bottles scrupulously clean. It will be found that Ewe Milk Replacer suits goats better than calf milk.

One Bottle Daily Through the First Winter

Whilst a six month-old kid will do perfectly well on hay, concentrates, greens, roots, beet pulp, etc., it will be decidedly beneficial to continue with one bottle daily until the spring. This is particularly important if the kid is to be mated early, at 12 months, instead of the usual 18 months, the goatling age.

FEEDING GOATLINGS

With the coming of spring, the last year's kid becomes a goatling at a year old, and she should soon be able to keep fit and grow well on the grazing, if it is good, and the area provided not too small. In different parts, the spring flush comes at different times, and although in the South-West of England there is usually ample grass to feed three goats on an acre by the end of April, or early May; farther north it may well be the end of May before there is enough grass to enable all pan-feeding to stop.

Goatlings do not need a milker's ration of balanced concentrates and many tend to get too fat. It is important for their future as breeding animals that they are kept lithe and fit. Broad bran in very small quantities, except on wet days, is the staple diet of the Author's goatlings. Bran is a balanced feed by itself and, containing wheat germ, is admirably suited to the goatling's needs. On wet days when confined, the youngsters may have any greenstuff which is available and, of course, hay.

AUTUMN BRINGS THE MATING SEASON

By August most of the nourishment is going out of the grass. The goatlings will begin coming into season any time from now, until October, when they should come in every three weeks regularly until mated. Gradually their rations should be increased and all forms of concentrates introduced, a few ounces at a time only, to accustom them to every kind of feed which they will be given as milkers. As their intake of dry food increases, so will their need for water; remember to always see that this is ready for them, warm and clean. Thus they will learn to drink after meals and upon returning to their pens, which is good training.

MINERALS

The mineral lick bricks in the goat houses may need renewing at this time of year, and the in-kid goatlings will require the extra teaspoonful of mixed minerals in the feed bowl once daily.

EIGHT WEEKS BEFORE KIDDING TIME

Now start to build up, or 'steam up', the rations so that by the week before the kids are due, the goatlings are getting 2 lb of concentrates, plus the sugar-beet feed, daily. This should allow them to start their production life with a modest 6–8 lb of milk and the feed should be increased as the milk yield goes up.

MAXIMUM CONCENTRATES

Most breeders with high-yielding goats find that they will not take more than 6 lb of concentrates a day but are able to give upwards of 15 lb (1½ gall), upon this, plus the other feeds, sugar beet pulp, greens, oatmeal drinks and treats from the house (such as stale bread rusked in a slow oven and potato peel done the same way). If the grazing is very good, the high-yielding goat may even have to be restricted so that she has room for her production ration. Herein lies the skill of feeding; and also in making sure that there are no set-backs to spoil the good milk record for the 365 day lactation, or the milking competitions which may be won at the summer shows.

ROUTINE DOSING

Because domestic goats are usually confined to a few acres it is essential to keep the worm count low or a vicious circle is set up with the goats becoming re-infected as soon as dosed.

In the wild state these animals remain very healthy, seldom returning immediately to pasture recently grazed, usually browsing rather than grazing and also being free to find for themselves the shrubs, trees and herbs which act as a vermifuge.

For routine dosing the reader should see 'internal parasites' in the chapter on illnesses. It is suggested that one dose of Nilverm be administered about three weeks after kidding, with an autumn dose of Thibenzole; any animal which is disturbed by the one medicine should be noted and given the other the next time. It may be necessary to dose more than twice annually; so much depends upon management.

Stall-fed goats are *not* immune to worms as some novice goat-keepers imagine to be the case, and an animal which has been kept in almost laboratory hygiene may still suffer with internal parasites.

Coccidiosis Treatment First

If *Coccidia* are known to be present, are suspected or are usually troublesome in the particular herd, it is wise to dose with Sulphamezathine first, then wait a week, and worm the goat. Again, see Chapter 21 on illness.

PEAK CONDITION AND YIELD FOR SHOWS

Most of what has already been said about feeding applies to the goats which are being prepared for exhibition, but a timely warning is also due, because it is all too easy to over-do the coming champion and so spoil her chances at the last minute. Although her every need, for maintenance, for milk, for general health and show bloom, will be attended to, never let her get over-fed so that she is not ready and eager for the next meal. If she is stall fed, always attend to her daily exercise. The goat is an active animal and should not be allowed to laze about and stuff. Her digestion, and her figure, will suffer.

Like the racehorse, she may be being brought to her peak for one special show, such as the Royal, where she will meet the very stiffest competition and, like the racehorse, the timing of this peak condition is all important. The Author does *not* hold with the use of injections and drugs to boost milk yield, but only those natural foods which are healthy and harmless.

BUTTERFATS

A perusal of the results of milking competitions in the British Goat Society's journal will enlighten the new exhibitor on the large number of disqualifications for low butterfat, particularly is this the case when the yield is high, and also it is more noticeable in some breeds than in others.

Many people have tried out different theories on how to achieve high butterfats, mostly without success. Without fail, some goats will produce excellent tests from 10 days until about 30 days after kidding, then gradually the butterfat content drops, to rise again in late summer, as the yield diminishes. Certainly poor condition does not help. Feeding extra protein such as raw eggs may offset the decline. The feeding of copious quantities of branches and very good hay, as opposed to lush greenstuff, is usually recommended.

RATIONS FOR MILKING GOATS

Summer mixtures, used also in Winter, with addition of third feed warm, as below. Rate of feeding, about 3½ lbs per gallon of milk *daily.*

Parts by percentage:

(1) 25% broad bran
25% dry sugar beet pulp
30% flaked maize
10% crushed oats
10% kibbled linseedcake

(2) 30% decorticated ground-nut cake
50% flaked maize
20% crushed oats

(3) 33% flaked maize
67% coarse dairy mixture (proprietary)

(4) 35% linseed cake
25% flaked maize
25% split beans
15% bran

(5) 25% soya bean meal
50% flaked maize
25% broad bran

(6) 48% dairy nuts
12% linseed cake
22% flaked maize
10% crushed oats
8% broad bran

WINTER MIXTURES

(1) 35% flaked maize
15% middlings
35% broad bran
15% soya bean meal
This mixture is fed damped (not wet) with hot water.

(2) 45% sugar beet pulp
45% flaked maize
10% broad bran
The sugar beet pulp (or sugar beet nuts) are soaked for four hours in boiling water then dried off with the maize and bran.

These two rations make a useful hot winter feed.

CHAPTER 13

HOUSING

GENERAL PRINCIPLES

The goat house may be a converted shed or garage, but the main essentials are that it should be light, airy, free from draughts and water-proof.

Plans for a female goat house, and for a male goat house, are given, the male house being purpose built and strongly constructed of concrete blocks on a concrete floor. Concrete blocks make excellent division walls when constructing pens in an existing building. The male goat may be managed in this type of house without contact with the attendant. The hay rack and feeding pan are attached to the door of the sleeping quarters and serviced from outside, as also is the water pail. See A, C and D, on plans.

The floors of both run and sleeping quarters are gently sloping to drain away to a 'soak-away' outside the building. The roof of the sleeping quarters is extended part way over the run.

SIZE

Male Goat House

The minimum size of the male's house is 6 ft square × 6 ft 3 in high at the front, and the run is 6 ft × 10 ft.

The height of the walls is extended with stout rails to 6 ft. Males do sometimes jump very high; therefore, it is not safe to have any male run lower than this.

Female Goat House

The plan illustrates a house which will hold four milkers comfortably. Each pen is a minimum of 4 ft 6 in × 6 ft, which allows enough space for kiddings (A on the plan).

In this plan, the pens are divided into two pairs with a communal hay rack (B) between each pair.

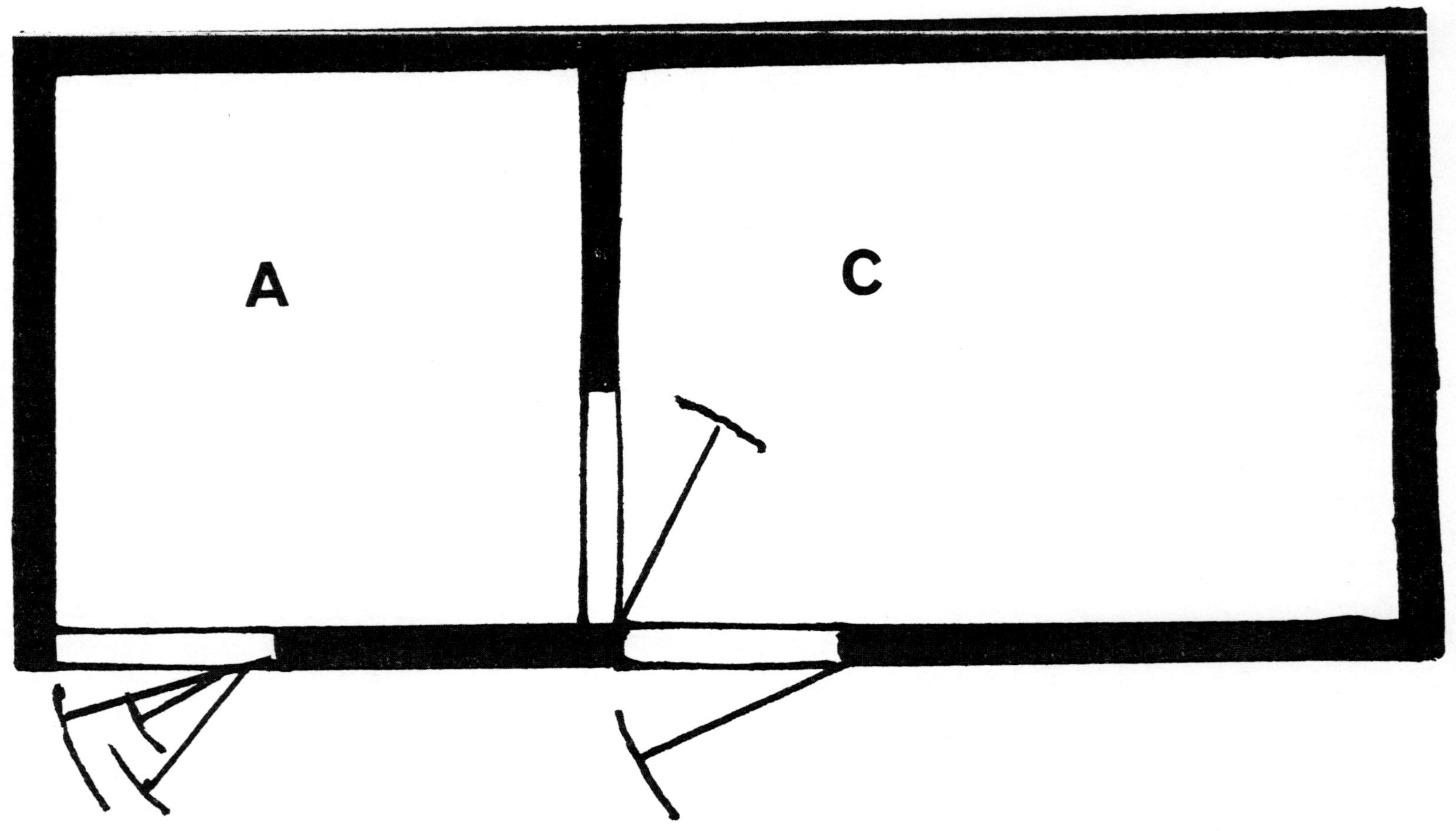

Diagram 13-1—Plan of Male Goat House (showing arrangement of doors to save unnecessary handling)

The floors slope in the direction of the arrows and a shallow gully runs along the outside of the pen fronts to the soakaway (D). The service passage must be wide enough to wheel a barrow down. The doors are designed to fasten open across the passage. The windows are large and on the south side.

Roofing can be of various kinds but, if of galvanised iron sheets, this should be insulated to prevent condensation.

Flooring might well have to be considered as temporary as, for example, for a tenant, not an owner of the property, and concrete might not be practical for this reason. In this case a well-trodden chalk or sand floor is quite possible to keep clean and free from odour, particularly if 2 in of dry sawdust or wood shavings are used under wheat straw.

On a temporary basis, it might be necessary to use hay nets instead of hay racks, but this is most undesirable for small kids and in this case it would be best to use the small portable hay racks normally used for shows.

A mineral lick brick holder would complete the essential furniture of both female and male goat house. Ideally, there would be a fodder store at one end of the building and a milking parlour with milking bench at the other, the whole being under one roof. In no circumstances must the male be housed alongside the female's pens.

Hot and cold water laid on, and an electric point for an infra-red lamp—so useful for small kids born in very cold weather, or for a sick goat—are refinements which few goatkeepers can install, but are nevertheless most desirable.

A NO-WASTE HAY RACK

The hay rack shown in the diagram is the best design for saving hay, of which goats are extremely wasteful if fed in an ordinary rack or hay-net. It is double-sided and makes a good division between two pens.

The basic plan is of a large box, 5 ft 6 in high × 3 ft wide and as long as may be required.

The outer wide-spaced rails prevent the animals from waving their heads about with a mouthful of hay; any which they do drop falls into the box below and is salvaged later to be fed to goatlings, males, or horses, calves and rabbits.

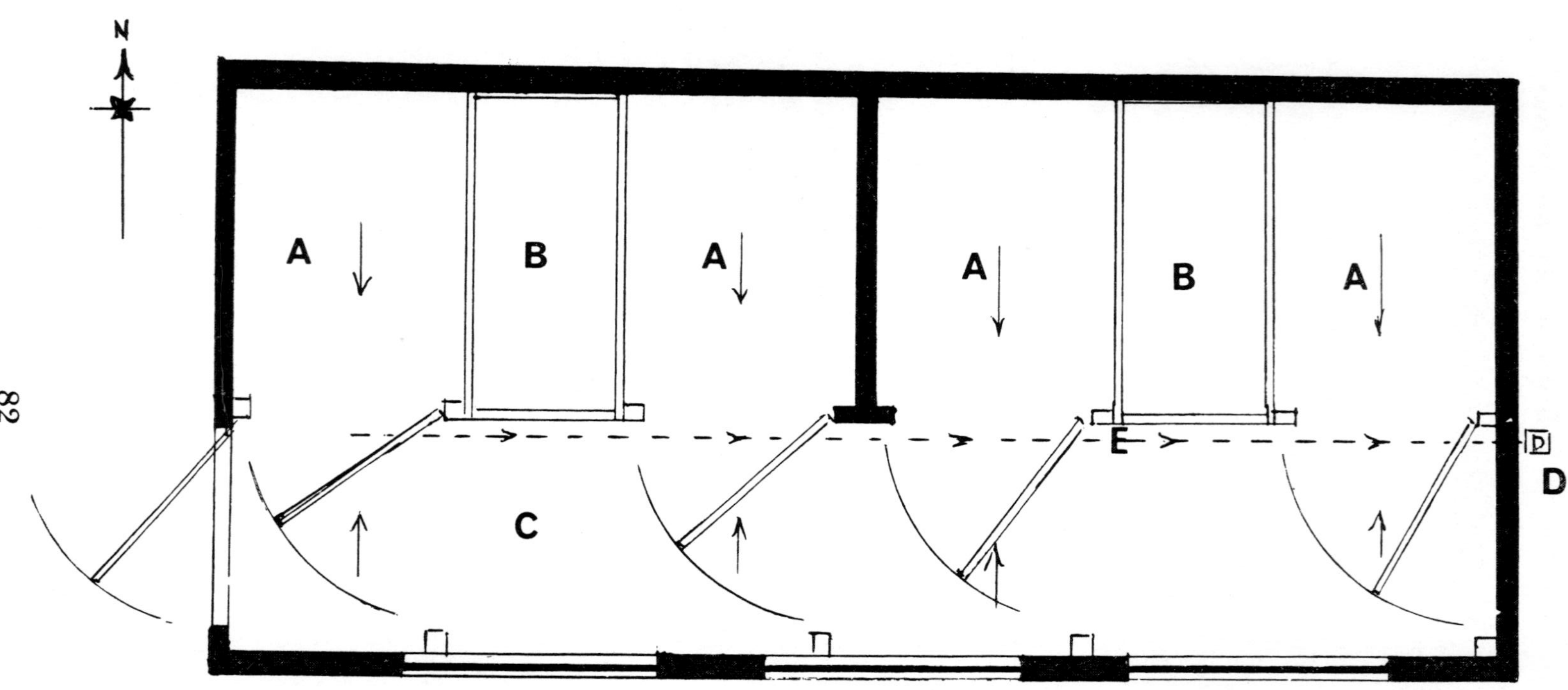

Diagram 13-2—Plan of Female Goat House

Construction

Make up the ends first, for which 2 in × 1 in frame is adequate, filled in with ply board, chipboard or plain boarding, but not hardboard which, in its normal thickness is not strong enough.

The ends measure 3 ft × 5 ft 6 in.

Next the sides; a frame of 2 in × 1 in timber, 5 ft 6 in high × whatever length is required for the rack. A 2 in × 1 in rail is fitted across the length of the frame, 2 ft 6 in from the lower side, this being filled in on one side only, thus making provision for two doors into the boxes, for extracting hay.

Next fit vertical spars above this at 6 in intervals, firmly securing to the top and centre rails. It is best to take the sharp edges off these spars.

Having made both ends and sides, firmly bolt them together and fit boards to the bottom just clear of the floor to keep the hay from becoming spoilt.

Across the top of the solid part of the box fit more 2 in × 1 in rails at 6 in intervals, narrow-side upwards. These will support the actual hay rack, which is made of two pieces of weld mesh 3 ft wide and the length of the rack. Use 3 or 4 in weld mesh—wire netting and chain-link fencing will not do.

The weld mesh is fitted into the rack in a V formation, with the bottom of the V secured to the centre of the cross rails and the other edge to the top-side rails.

The rack is now ready for use, and one about 6 ft long would hold two whole hay bales, but it is preferable to put less hay in and more frequently, unless the attendant has to be absent for several hours, as the goats will eat it better if fluffed out.

FIELD SHELTERS

The ideal at which to aim is a small open-fronted shelter in each field so that the slightest shower does not bring all the goats hurrying back to base.

These need not be elaborate and, in fact, an old pig arc works very well—about seven goats can crowd into an arc of standard size.

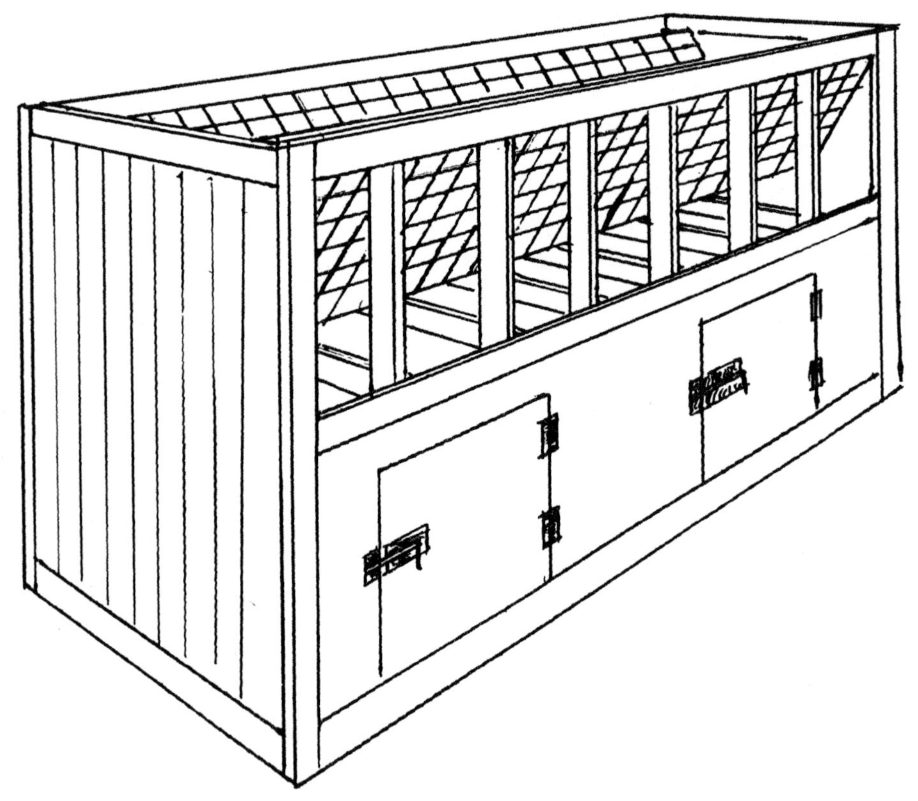

Diagram 13-3—A No-waste Hay Rack

CHAPTER 14

BREEDING

THE SEASON

The female goat comes into season every three weeks from the end of August to the middle of March, though those without a male on the premises will probably not notice the first signs of oestrus until October.

Age at Which to Mate

Goatlings, which are animals which have not kidded and are over one, but under two, years old, are usually mated during their second autumn, when they are between 15 and 18 months old, so that they will kid in the spring of their second year, when about two years old.

It is sometimes practical to mate a very well grown and early spring-born kid when she is around 12 months of age, for then she will prove extremely useful as a winter milker, kidding about July when the other goats are beginning to drop in yield. It must be emphasised, however, that a kid to be mated this early *must* be well grown, sound and in excellent condition; the following year she should not be mated, but be allowed to grow on whilst she milks (this is termed 'running through').

Mating

The signs of the goat being in season and ready for mating are more noticeable when a male is kept on the premises, but if there is no male and the owner is uncertain, a rag rubbed well over the male goat's head and kept in a tin, will often produce signs of oestrus when presented to the female.

She will wag her tail in a rhythmic manner, quite unlike the occasional flick, and may be slightly swollen and pink under the tail, with a slight clear discharge. Some goats are very noisy and bleat a lot, but others remain quiet.

If the female has to be taken some distance by van to the stud, she should be steadily driven and, on arrival, she should be taken out and allowed to rest in a pen, with a little hay to settle her down. A few minutes later, if she is ready, she will stand for service, which she might not do if taken to the male immediately upon unloading.

A second service may be desirable if the first was not a very satisfactory one, and in this case the male should be taken back to his pen for 10 minutes rest. If there is no reason to suppose that the first service was unsatisfactory, there is really no need for another. A young male kid, giving his first service, is usually allowed another, because it sometimes happens that the first is ineffective.

Certificate of Service

The owner of the female being mated should have with him the goat's registration number, which is shown on her pink card, as the stud goat owner will need this number for filling in the certificate of service, which he will give after the service when the fee is paid. The female's owner will not be able to register the resultant kids without this document, so it should be carefully filed.

If the female comes into season again, three weeks after service, the owner is allowed one free return service and should notify the stud owner at once. Occasionally, a goat will pass over the third week, uneventfully, and then return at the sixth week.

SIGNS OF PREGNANCY

If the goat is only going to give birth to one kid, it may be almost impossible right up to the last to be certain that she is in-kid, but if the other signs—a tendency to want to go dry, a good appetite and no further coming into season—are apparent, the goat should gradually be dried off after the first two and a half months so that, by the time she is three months in-kid, she is dry or thereabouts. A good milker which is due to kid in spring and which is not given sufficient time to dry off, may well begin to rise in yield with the lighter evenings and fresh greenstuff. If this should happen, her subsequent lactation will suffer and her kid will have no colostrum, for she will still be giving normal milk at kidding-time. Liquid medicinal paraffin will clear its bowels of meconium, the first motion, but will not give the little animal that vital dose of antibodies with which it staves off harmful bacteria during the first days of life.

To dry off the too willing milker, leave about a pint of milk in the udder at milking time and reduce the concentrates gradually. Milking at uneven times also helps, because the goat is taken by surprise and fails to let down her milk. When the yield is down to a couple of pints daily, reduce milking times to once daily. Finally, milk only every other day, then every third day, until at eight weeks before the kids are due, the goat is dry.

FEEDING BEFORE KIDDING

Assuming that the goat is now dry, begin to build up her ration again, ounce by ounce, until she is taking 2 lb daily of a balanced milk production ration. Her kale, root and hay intake will depend largely upon her size. Hay should never be rationed and a mineral lick *and* mineral additive are essential at this time.

WORMING BEFORE KIDDING

Although modern worm medicines are far less dangerous than the old ones, they are, nevertheless, a drug, and should not be given when the goat is heavy in-kid, or immediately after kidding. About six weeks after mating is late enough, and then the goat will require another dose about three weeks after kidding.

Goats vary very greatly as to the length of time before kidding that they begin to make an udder. A 'maiden' milker—that is, a goatling which produces milk before she has been mated and consequently has to be milked perhaps just once a day—may have a much more mature-looking udder than one which has not milked, and the latter may not spring an udder until after she has kidded. However, this need not cause concern that she will be a poor milker or, conversely, that the maiden milker will be a particularly good milker.

An adult goat which has kidded before usually shows increase in her udder about three weeks before her kidding date. An over zealous milker may have to be eased before she kids, but only if the udder is uncomfortably tight and shiny—for the same rule applies—the milk will be normal milk by the time the kids arrive and they will miss out on their colostrum. It is much better not to touch the udder if it can safely be left.

KIDDING PREPARATIONS

Since a week either way is not unusual, be prepared for the kidding a clear week before the event is due and do not be alarmed, providing the goat seems well, if she is up to one week late.

The box must be clean, with a good bed, but not enough straw for the kids to become buried in it. All buckets, including the water bucket, should be removed after use, for a kid can be dropped into a water bucket. The goat must not be tied, and any other animal with whom she normally associates is better removed, for her best friends will not be tolerated at this time, although most goats find comfort in the quiet presence of their owner.

SIGNS OF IMMINENT KIDDING

Usually, the goat which is going to kid during the day will not willingly leave her box when the others go out to graze. She will seem restless, apprehensive, may bleat and will eat a little hay, then scratch her bedding about and lie down for a while—getting up again and repeating the process—until finally, the second stage of labour approaches and she begins to stretch. By this time, her shape will have altered noticeably, the tummy will have dropped and hollows will appear on each side of the tail which seems to stick out along a bony ridge. There may be a white discharge, followed by a slimy mucous which in turn is followed by the water bag.

If the kids are correctly presented for easy birth the nose and fore hooves of the first kid will appear within the water bag. On no account touch the bag, but allow the goat to proceed naturally if all is well. She may groan or shout as she strains, but this is no cause for alarm, providing that she is making progress, and at this stage, with nose and feet in sight, a few good heaves and the kid should be born. She will probably rise to her feet and, in turning to lick the kid and clean it, twist and break the cord naturally. A soft paper kitchen towel for drying the kid and wiping out its mouth is useful to have by in case a second and even a third kid arrive so quickly that the mother has not had time to deal with the first. It is also a good idea to have handy a tea chest or large basket in which a warm bottle is placed, covered with hay, and into this the first kid may by put. This is a precaution because in the event of a multiple birth somewhat disturbing the mother, she will sometimes tread round and round on the first one or two, whilst contemplating the birth of two or three more!

HELP

The goat is a hardy, strong animal by nature, and usually manages her kidding unaided, but it is most unwise to leave her for too long without making sure of her progress, for it can happen that a dead kid is impeding the process, or that a wrongly presented, or very large kid, cannot be born without help.

As a general rule, a goat which has strained regularly for about 45 minutes without any sight of a kid in its water bag, must give rise to suspicion, and if the attendant is not capable of making an internal examination himself, a good shepherd or goat keeper or a veterinary surgeon should be called in within the hour of the first really hard regular straining having begun.

An experienced goat keeper would, at this stage, scrub up, having cut back the finger nails as short as possible and, using some obstetric cream or soft plain soap to lubricate the hands, would insert one hand with the fingers all together in a tight cluster, very slowly and gently. Sometimes the trouble can be found at once for it may be a kid has one foreleg back, and this may be straightened out and brought to lie alongside the nose with the other foreleg. Make absolutely certain that both legs *are* forelegs and both belong to the same kid, before taking action. **In other words, proceed with great care.**

If the kid is presented bottom first, or hind legs first, the birth usually takes longer, but it will usually be completed on its own.

A very big kid with correct presentation may need help, and this must be by means of pulling on the two front legs in a downward direction towards the mother's hocks, *only* when she strains. Relax when she relaxes.

When the actual kidding is completed, the afterbirth or placenta should be discharged within a few hours, **and on no account must this be pulled.** There should be one for each kid, except in the case of identical twins.

An assisted birth is far more likely to be followed by a retained afterbirth, than is a natural birth, and this possibility must be carefully watched for as the goat closes up after about 10 hours and, unlike the cow, the afterbirth cannot be removed manually later on.

Any suspicion that all is not away should be followed by an immediate antibiotic injection and the goat's temperature should be taken to make sure that there is no infection. (It should be 102–103°F.) The goat should not be allowed to eat several afterbirths completely, because this can give rise to seious indigestion; however, for her to consume one, or part of one, is perfectly natural.

Drink

After kidding those goats which will not drink oatmeal gruel (made by pouring boiling water over a paste of medium oatmeal mixed with a little cold water and a teaspoonful of salt), will sometimes enjoy the warm water and molasses from sugar-beet-pulp soaking.

A warm bran mash may also be given at this time. This is made by pouring boiling water over about two double handfuls of broad bran, with a little salt, and leaving the bucket covered for about one hour or until cool enough to feed. The hay rack should be emptied and replenished with a little of the best hay.

A warm wash-down of her hind parts will make the goat more comfortable before she lies down to rest.

The kids should soon be up and looking for their milk and if their mother has a large, low, udder, it may be necessary to assist them and make sure that they have found her teats. A weakly kid, which has perhaps had a delayed birth or may be the smallest of the family, is best given a small bottle feed. Pour hot water into a baby feeding bottle and milk a little of the goat's colostrum into it immediately after tipping the water out. Colostrum does not heat up without solidifying unless the bottle containing it is placed in a container of hot water. A Maw's Baby Bottle has a wide enough aperture to allow milking directly into it.

FREEZE OR FEED BACK SPARE COLOSTRUM

When several goats are due to kid, it is a safeguard to have a little spare colostrum deep-frozen for future use in an emergency, but if it is not wanted for kids it will benefit the mother to allow her to drink any spare for the first two or three days.

AFTER-KIDDING DISCHARGE

There will be a bloodstained discharge for two or three weeks after kidding. This is normal and will require washing off. It does not begin until a few days after kidding.

FEEDING—FIRST WEEK

The freshly kidded goat should not be fed normal quantities of concentrates for a few days after kidding. Suitable foods are the best hay, green food according to season, but leafy branches if possible and light feeds such as oven-rusked stale bread and bran mashes (if liked). Take her gradually back to the balanced milk production ration, first using soaked beet pulp or plain flaked maize. Trying to push her at first may result in her refusing food and it is important for her future milk production that her intake steadily increases.

PEAK YIELD

In some breeds, the highest daily yield of milk is reached about three or four weeks after kidding; in others, it may take three months to reach the peak. Fortunately, those goats which take longest to go up are very often the longest to drop in yield.

In some breeds, the highest yield may be given in a goat's first lactation; in others, notably the pure Toggenburg, her lactation yield may still be rising in her fifth lactation, when about 7 years of age.

AFTER-KIDDING ROUTINE

About three weeks after her kids are born, the goat should have her routine worm dose and her feet—which may not have been trimmed for some two months whilst she was heavy in-kid—will need careful trimming (*see* 'Work in the Goat House').

Examine the New Born Kids

So far, the goat herself and her needs at kidding time and after, have been dealt with, but there is one very important point which must never be overlooked. This is that the breeder must always take each kid into a good light and examine the tiny teats of both male and female kids. This is because faulty teats are very hereditary and although a small supernumerary teat alongside the normal one may easily be removed if it has no duct, the fact remains that by so doing, the breeder may be storing up future trouble in his strain. The progeny, or their progeny, may give birth to or beget kids which have a fish-tailed teat with two ducts, which is impossible to milk, or a double teat with two ducts, too near together to be practical to handle. Any of these faults would eliminate the animal from any chance of success in the show ring.

Extra Teat  *Fishtailed Teat*

Diagram 14-1—Abnormal Teats

Cull and Produce a Sound Line

Any hereditary fault must be dealt with ruthlessly and the kid destroyed at birth. In this way, and only in this way, will the breeder arrive at stock which is one hundred per cent sound and reliable. It takes only two outcrosses to stock which has not been so carefully bred for trouble of this kind to recur.

TRUTH FOR PROGRESS IN BREEDING

It is an unfortunate fact that few breeders of any form of livestock will say quite honestly what are, or have been, their own particular difficulties in breeding. Recently a line which had been kept clear of teat troubles for 40 years by one breeder, was outcrossed twice, and on the second occasion a faulty teat was found on the kid. This fault must have been present in both outcrosses, probably as a carried factor and not apparent in either stud males used. Previously this breeder had used only his own males and had closely line-bred and in-bred his own faultless stock, culling when necessary.

Conversely, the importance of knowing the strain over many years cannot be over-emphasised, for only in this way can one safely in-breed without accentuating any faults of any kind whatsoever.

Many novices fail to establish the required type, or to come up to the exhibition standard at which they hoped to aim, because they have the mistaken idea that they must outcross as far as possible, and they, therefore, search for a proven male which contains none, or little, of their own goat's breeding in his pedigree. A lucky fluke may result, but this will be a 'one off' process, and there will be no consistancy.

Admittedly, it is much more difficult to achieve all-round success (type, milk, butterfats and long lactation), in some breeds than in others, for in some the 'spade work' has already been done for us by clever breeders. But this uncertainty and the problems which crop up in one's breeding plans, all add to the interest, and the satisfaction, when one breeds a champion who herself breeds a champion.

INTER-SEXED KIDS, AND OTHER FAULTS

The so called hermaphrodite kid, which has neither the complete reproductive organs of the male, or of the female, and the sterile male kid, which are linked with the state of natural hornlessness, will, of course, cancel themselves out, and will not breed at all, although they may be a source of great disappointment. Here the rule must be, that a naturally hornless female must always be mated with a disbudded male, or *vice versa,* taking care that at least one parent is disbudded. There is about one chance in a thousand of a horned hermaphrodite turning up, but all kids should be examined at birth. Determine within a few hours of birth whether the kids are hornless or will be horned (*see* chapter on Disbudding).

Sometimes when there has been a delayed, or crowded birth, a kid will seem stiff in the pasterns and be unable to place its feet down flat on the floor, whilst the knees remain bent. Occasionally the ears in prick-eared breeds are bent or flat at birth. These conditions should right themselves within a few days.

Feeding the Orphan ?

CHAPTER 15

SHOWING GOATS

PROFITABILITY

Exhibiting goats is not likely to be profitable, for the entry fees are usually related to the prize money offered and if the cost of the transport be added to the former, with possibly the addition of money paid out for temporary help at home whilst the owner is away showing, there may not be much direct monetary gain. However, the indirect 'profit' may be enormous for winning animals should add considerable value to their stock, and lustre to the herd. Moreover, the owner will have fun and learn.

SHOW SCHEDULES

The British Goat Society's monthly journal prints a list of all recognised shows (shows which comply with the Society's regulations and for which all exhibits must be registered). From this the newcomer to goatkeeping may choose the nearest and most suitable show at which to start, or he may decide to begin with the local affiliated goat club's own young stock show. This schedule will be obtainable from the local club secretary.

Having obtained the schedules, make a careful study of the classes and the qualifications; also make sure, on looking up your goat's registration card, that you are entering her in her correct class.

CLASSES

At the larger shows, there are usually separate classes for each breed. At the smaller, some breeds may be grouped together as, for example, Saanens and British Saanens, Toggenburgs and British Toggenburgs. The class for **Any Other Variety** goats means all British Herd Book animals and any Breed Section goats for which no separate class has been provided. Sometimes there is a proviso that if four or more of any breed in the AOV class are forthcoming a separate class will be added. At a club show there may be classes divided into age groups only for Milkers, Goatlings and Kids.

MILKING TRIALS

Most Agricultural shows with a Dairy Goat Section hold milking competitions which are based on the 24-hour yield of the entrants. These require the presence of the entrants at a set time the afternoon before the show is judged, usually around 5.30 p.m., so that the milkers are stripped out by the stewards after the evening milking at about 6 p.m. The milk of the next 24 hours produced at the show is then weighed at each of the two milkings and samples then taken for butterfat testing. The milk must not be less than 5½ lb when weighed over 24 hours, or the goat will be disqualified, and the butterfats must not be lower than 3% for the same reason. The points required for winning a star or QX in the milking competition are given in chapter 16, as also are points needed for the retention of a *Breed Challenge Certificate* or *Open Challenge Certificate* which go towards the making of a breed champion and a (full) Champion respectively. These may be awarded, then lost on test; an explanation of the somewhat complicated details needs to be studied. When reading the schedule, note that the milking competitions are sometimes divided into breeds and sometimes only into one each, for first kidders and second and subsequent kidders.

Having studied the schedule and made the entries carefully, filling in all the information required—such as class numbers, date of birth of entrant, date of last kidding, herd book numbers, sire and dam—be sure to post in good time, for these entries vary considerably as to closing dates (stated in schedules).

HANDLING

Assuming that management is good and that the goats are in good condition, from the time the first schedule is posted a little show training daily should be the aim. A few minutes spent walking on a loose lead, in a circle, and out and back, then standing in such a position that all the goat's good points are shown to advantage, will make handling on the day so much better and easier.

An example of a young goat standing well and naturally is given in Figure 8, looking alert and not as if strung up by the handler, gives some idea of what is required. Contact is kept by placing the forefinger of the hand holding the lead against the goat's neck, rather in the same way that the well-trained horse 'listens' when the reins are taken up. The goat soon learns that she is expected to stand still in just that position as long as the finger is against her neck. Movement away from the desired perfect poise may be caused by boredom if the judge is slow or if the class is a very big one, so do not expect the animal to stand whilst he is away down the line. But be prepared, and watch the judge so that you may take up the position before he again looks your way.

Young stock new to the ring also fidget and must learn to reassume their stance with the aid of a toe gently tapping a foot out of place, or a slight movement from the collar, backwards or forwards, as the case may require.

Never shove the goat around by hand on its quarters because this only causes your goat to hunch herself up, with all four feet in a huddle under her with too much slope to her rump. Another fault of handling, often seen, is the exhibitor who clutches the goat's head much too high, thereby giving the back-line a dip or accentuating a small existing dip beyond hope. *Look at your exhibit from all angles.*

Examination by Strangers

Practise your showing with a friend if you can, getting the goat used to having her mouth examined, her feet picked up and, if a milker, her udder felt. Some goats work themselves into such a frenzy of nervous tension that they ruin their own chances by refusing to stand nicely after being examined.

SHOW PREPARATION

The day before starting off to attend a show most white goats will need a thorough shampoo. Swiss marked animals will probably look better for a wash of the white markings.

Figure 8—Unobtrusive control

(The goat stands perfectly but relaxed—with the handler's contact on her neck)

Any good shampoo will serve; horse shampoo is cheaper than human's. First wet the goat with warm water, then rub in the shampoo. A small brush is useful for the legs and a sponge for head and under the tail. Rinse in warm water and repeat, using plenty of water for the second and final rinse. Squeeze out surplus moisture, rub down with a dry towel and then brush the hair flat.

Rug the goat—if very thick-coated—with a small towel under the rug, which can be removed after a few minutes. If it is possible to walk the goat about until she is dry, so much the better; if not, return her to a clean bedded box and later change the rug for a light one to keep her clean overnight and on the journey. Instructions for making goat rugs will be found, along with show lead making, in Chapter 17 which covers accessories.

Feet

The exhibit's feet should be trimmed several days before the show because there is always the risk of laming the goat through cutting too deeply or because the goat struggled and pulled.

A small amount of tidying-up with hair clippers or surgical scissors may be helpful towards a smooth outline, particularly around the udder and the beard, if any. Sometimes there is a thick pad of hair under the brisket, which looks untidy from a side view. Do not remove the hair from inside the goat's ears—it is there to keep dust, etc., out.

Packing for the Show

Your own needs should be packed into two containers. The personal 'wardrobe', including white overall, thick jerseys, a change of socks, wellingtons and mackintosh, will go into a kit bag or rucksack. Kitchen equipment and food can be placed in a tuck box or tin. The goat's concentrates are best packed separately, the small amounts of each mix then placed in a large polythene bag or small bin.

Hay is also best in a large polythene bag; it is a wise precaution to cut the bale and ensure that it is a good one, before putting the wads into the bag.

Branches and greenstuff can be packed in paper sacks and tied in two places to flatten for carrying on the roof rack. If the show is of more than one day's duration, crops like lucerne and kale benefit from wet newspaper wrapped round the stalks first.

Usually, bedding straw is provided and the marquee contains separate pens—although not always one for each exhibitor's goats separately, doubling-up being usual for goatlings and for kids.

Some exhibitors prefer to sleep in their vans or cars, whilst others like to stay alongside their goats and put up a camp bed, or put down a ground sheet, in a spare pen. Whichever way is chosen, a sleeping bag and an extra blanket or two, plus a pillow, will be needed.

The bigger agricultural shows usually provide some form of kitchen, where there may be a gas ring for the use of exhibitors. When one is taken, it is usually placed upon a trestle table, alongside the others. The wisdom of this is doubtful, for although the thinking behind it is that the gas cylinder in the spare pen is unsafe, there must be more danger if one faulty gas cylinder ignites a dozen or more. All shows should have a fire extinguisher where there is hay or straw; it says much for the careful use of show facilities that there are so few fires.

Catering

Do not rely upon there being a herdsmen's canteen, although this is often the case at the big shows. Aim at being entirely self-supporting for the duration of the show.

Besides all the obvious requirements a torch should be packed as well as grooming equipment, the show envelope with numbers and instructions on times of arrival, stripping out and judging. Also take hay racks and/or hay nets, extra string and a small sheet for making the pen cosy if the weather is cold. Some exhibitors have a long narrow light-weight tarpaulin to run the length of the pen fronts at night.

Secateurs for cutting branches will be needed, for it is astonishing how much the goats will devour, when entirely indoors, and most exhibitors go out during the evening to stock up for judging day. Each goat will require a feeding pan or pail, and a water bucket. Each goat in milk will need a milking bucket. Rugs should also be packed, at least for each milker, and one small size mineral lick brick can be passed around. First Aid equipment is advisable too, as well as bottles and teats for kids.

TRAVELLING

Goats should not be loaded for two hours after a full feed, but they do not seem to be upset if taken direct from their field to the van. Speed will depend largely upon the type of vehicle, a smooth journey being the necessity—around 40 m.p.h. is usually fast enough and does not upset the occupants of trailer or van.

On arrival, having found your pens, straw them down and fill the hay nets or racks and water buckets. The goats should start to eat hay at once and should not be fed anything else for a while, the next item on their menu being branches. They will usually sit and cud quietly after a time, but if there is a bad traveller amongst the party, a dose of MacLeans Powder will usually put her right. A more severely upset goat will generally be improved by the administration of a bottle of Guinness!

At about 6 p.m., after the owner has milked them, the goats in milk will be stripped out by the steward (they are then left until after the preliminary udder inspection between 6 a.m. and 7 a.m. the next morning).

By now they should be settled and ready for their normal evening concentrate food. Do not try to stuff them, but give exactly the amount and the type of concentrate which they are used to having at home. It often happens that a heavy milker will turn up her nose at her erstwhile favourite dish. If this occurs, change it by all means to something she fancies, but do not leave her with *both*. Goats which at home love soaked sugar-beet pulp may refuse it at shows, and *vice versa*.

Before settling her down for the night, the milker is usually rugged, given fresh water and a nice large bundle of branches tied to her pen. Her hay rack is replenished, leavings being emptied and fed to goatlings.

Goatlings and kids at shows must not be overfed and are therefore not usually penned next to the cosseted milker. Put the spare pen between them, so that they cannot stretch over and steal all her greens.

It should not be necessary to wash any goat if brought clean in a clean van, but a quick wipe with a sponge early in the morning may be needed, before going into the ring.

Figure 9—Anglo-Nubian male kid Spean Truffle

(Bred by Joan Shields, owned by Mrs L. Partridge, Wales)

JUDGING MORNING

An alarm clock is seldom needed on the big day because exhibitors begin stirring at 5.30 or 6 a.m., and cups of tea are brewed. The goats are given fresh hay, water and a few branches to keep them busy until required in the ring.

The milkers are groomed, white overalls and numbers found, the milking buckets have the numbers attached and are turned upside-down outside the pens ready for action when all the milkers' classes have been into the ring.

The milkers usually go into the ring in catalogue class order and stand with their backs to the judge, in numerical order. The judge will make a thorough examination of the full udder, will probably then have the exhibits standing sideways and make a quick note or sketch of udder shape and teats. He may, or may not, place the animals and say, "Come back in that order."

When all the milkers' classes have been in, the goats are milked and the buckets collected by the steward, who begins to weigh and take butterfat samples as soon as all the milk is handed in. It is interesting to write down all the weights against the goats' numbers, as well as just your own, for comparison later.

Once the milk-weighing is over and the goats fed, the exhibitor's own bedding can be put away and breakfast taken, when there is usually about an hour before the judging proper begins at about 10 a.m.

THE CHAMPION MILKER

The milkers go back into the ring in class order and the awards are made. When all the milkers' classes are judged, the First, Second, Third and Reserve (or Fourth and Reserve)—in fact, all goats awarded a prize and the Reserve—go back into the ring together for the inspection/production line-up.

This really amounts to inter-breed judging along with the points later awarded on milk and butterfat and time since last kidded. The goat heading the line-up may be the **Best Goat in Milk,** but her owner will not know if she has won the *Challenge Certificate* and the *Inspection/Production Challenge Certificate,* until all the results are worked out.

BREED CHALLENGE CERTIFICATE

The Judge will usually mark his line-up, starting with half a point at the bottom of the line, and finishing with the goat at the top of the line having the most points, half for every goat below her.

He will then decide whether to award the *Breed Challenge Certificates* to those animals at the head of their respective breeds, not necessarily at the head of the class, for there may be more than one breed in a class.

Having thus completed his judging of all the milking goats, he will begin on the goatlings which, of course, he has not seen as yet. If the entry is a big one the kids may not be judged until the afternoon. Then he will judge the inter-breed champion goatling—and the same for the kids, perhaps calling for all Firsts and Seconds, or perhaps for only First prize winners.

THE SUPREME CHAMPION

Now comes the big moment when the champion goat in milk, the Reserve champion goat in milk, the champion goatling and the champion kid, are lined up for the award of the *British Goat Society's Rosette* for the **Best Goat In The Show.**

CLUB AND LOCAL AWARDS; CUPS AND SPECIALS

There follows the local awards, sometimes special awards for Novices, and often groups of three, owned and bred by one exhibitor, or mother and daughters. All are of interest and instructive to watch.

FINAL MILKING

The second milking within the 24 hours of the milking test will have to be completed by the time at which the goats were stripped out the evening before. Then the procedure is the same as for the morning milking and, when this is completed, the long day is over, the packing-up is done and the goats loaded for the homeward journey.

Spare a thank you for the stewards who have worked hard for no reward, but the success of the show and your comfort.

If returning home cardless, just think how much has been learned, and resolve to go again and do better next time in the light of that knowledge.

SOME ODD HINTS

Travelling

If your goat has to travel in the family car, take out the back seat and in its place lay a sack containing a small amount of sawdust or peat, over which can be put some paper sacks for extra protection.

Then, before inviting the goat to jump in, get the family smoker, if there is one, to blow smoke in her face. She will then 'oblige' before loading and will probably be clean in the car, so long as she is removed at once when the car is pulled up.

Leading Goats

When pulling a reluctant goat along on a lead, do not be astonished if she suddenly quivers and collapses in a shivering almost unconcious heap at your feet, but let go *at once,* loosening the collar and/or lead. You have put pressure on a nerve and quite literally knocked her out. She will soon recover if the tension is released.

Tying-up Goats

Although goats should never, never be tethered out, there are certain occasions when they must temporarily be tied up, as, for example, when several are housed in one large pen at feeding time. Make sure that the lead cannot go round and round any obstacle, thus causing the collar end to twist and throttle the goat.

Another cause of tragedy is a long lead enabling the goat to lie down and get the lead behind her hind legs; when she tries to rise it then gets twisted round them. This has the effect of strangulation through her own back legs tightening the lead the more she struggles.

Companions

CHAPTER 16

SHOW AND MILK RECORDING AWARDS

Prefixes and Affixes Awarded by the British Goat Society (at recognised shows and for milk recording)

SHOW AWARDS

A *Champion* female goat is one which has been awarded three open (inter-breed) *Challenge Certificates* for best milker in show, under three different judges, and also three *Inspection-Production Certificates* awarded jointly for points gained in the inspection line up, in conjunction with points gained in the milking competition. To hold a *Challenge Certificate* the female goat must have gained a total of 18 points in the milking competition.

She must also have been awarded a **Q*** (*see* below).

A **Champion** male goat is one which has been awarded three *Challenge Certificates* for best male goat in show over one year old, under three different judges.

A **Breed Champion** female goat is one which has won five *Breed Challenge Certificates* under three different judges and has also been awarded **a star** or a **Q*.** To hold a breed challenge certificate awarded for best of breed, a goat must obtain 16 points in the milking competition.

A breed champion male goat must win four *Breed Challenge Certificates* for best of breed, under three different judges.

A **Star** is awarded to a female goat which gains a minimum of 18 points in a recognised 24 hour milking competition, with a butterfat percentage of not less than 3·25% at each milking.

A **Q*** is awarded to a female goat in a recognised milking competition which gains a minimum of 20 points, and whose butterfat is not less than 4% at each milking.

The milking competition results are based on time since last kidding, pounds of milk given and butterfat percentages on the *morning* and the *afternoon* milk tests.

Figure 10—British Toggenberg and Anglo-Nubian goatlings

(In the Balaams Herd, browsing in the Trent Valley)

For each one pound of milk 1 point is awarded.

For each complete 10 days since last kidding one tenth of a point is given.

The maximum time points allowed are 3·6.

For each $\frac{1}{4}$ lb of butter fat 5 points are awarded.

At some very large agricultural shows 'solids not fats' are also taken into account.

No goat giving less than 5½ lb of milk in the 24 hours, or less than 3% butterfats at each of the two milkings, is eligible to compete and will be disqualified.

To enter the milking competition, the goat must have kidded a minimum of 10 days and a maximum of two years before the milking competition.

For every female goat in direct line, which has won a **star** or a **Q***, a numeral is added, thus **Q*4** indicates five generations in the female line possessing this award.

AWARDS FOR RECORDED MILK YIELDS

National Milk Records, administered for the British Goat Society by the Milk Marketing Board (include the recording of goat herds, some 46 herds being recorded in 1975).

The **R** prefix is awarded to the female goat which yields a minimum of 2000 lb of milk in a 365 day lactation. The two numbers after the R signify the thousands and hundreds; as, for example, R45 means that the goat has officially recorded 4500 lb in 365 days.

The **RM** prefix signifies that the goat has attained the *Register of Merit,* because her dam qualified for an R29, or more, as well as the goat herself.

The *Advanced Register* is for goats which have qualified for the R35, or higher, and have an average butterfat of not less than 3·5%. The dam must also have qualified for the Advanced Register, and the goat's sire must be out of a goat also registered in the Advanced Register.

MALE GOAT AWARDS ON MILK RECORDS OF DAM AND SIRE'S DAM

The **Dagger** sign (†) is awarded to the male goat whose dam and sire's dam have both won either a Star or a Q*.

The **Section Mark** (§) is awarded to the male goat whose dam and sire's dam are both **R** goats. The award will read 'Section 35/54', shown as §35/54.

The **Double Section Mark** (§§) male goat is one whose dam and sire's dam have the **RM** or **AR** prefix, and whose sire has the Section Mark or Double Section Mark. The Double Section Mark is followed by numbers indicating the amount of the yields, as with the Section Mark.

SM (Sire of Merit) indicates a male who has sired a minimum of five Q*, * or R milkers.

CURRENT RECORDS

Readers wishing to obtain information on recent milk yields should contact the British Goat Society.

CHAPTER 17

ROUTINE WORK IN THE GOAT HOUSE

DISBUDDING KIDS

Kids which will grow horns are easily distinguished from those that are naturally hornless by any seasoned breeder, but it will probably be a good idea for the novice to ask to be shown a hornless kid, and a horned kid, at a few hours old, so that he may learn what to expect.

A horned kid is very often the best of twins, one of which is hornless, and many breeders prefer to have disbudded rather than naturally hornless animals because of the sex-linked troubles which arise in an entirely hornless herd (*see* the chapter on Breeding).

Horn Buds or 'Knuckles'

The *horned* kid will have a spiral curl around a tiny bare spot where each horn bud will break through after a few days. If wetted this curl becomes more obvious.

The *hornless* kid will have normal hair covering a knuckle or bump, which is roundly smooth and grows on as a knob, rather than as a point.

The horned kid must be disbudded at between two and seven days; two days if a male, and usually four days if a female. Sometimes a premature, or weakly and small female kid may be left for about 10 days, but it will take the operator longer to do the job and in any case should not be left until the horns are through.

Some veterinary surgeons believe in leaving the kid until it is about three weeks old, then, using a local anaesthetic, digging the horn buds out and leaving quite a hole. There is no advantage in this method, but the very real risk of hitting the wrong spot with the needle and causing meningitis. This has resulted in several deaths and is not recommended.

A skilled breeder with much experience of disbudding will do this job every spring as the kids come along and should be asked to demonstrate the method for the benefit of the beginner, before the latter tries his hand at it. It is not

Diagram 17-1—Disbudding Kids Using Hot Irons

in any way complicated or dangerous; it simply requires a steady hand and a good holder for the kid. The disbudding iron must be one made for goats and must be cherry-red hot.

Good Fire + Two Hot Irons

Any fire into which the disbudding iron may be inserted until it is really red hot will do; an electrically heated iron such as is commonly used for disbudding calves has not been found very satisfactory.

Two irons are better than one to ensure that each is really ready for use. The first application of the iron cauterises and deadens any further pain and is less painful than a local anaesthetic. The kid will suffer no after-effects and will be ready for its bottle immediately it is finished and put down.

The Method

The kid's holder is playing the most vital part in the operation, for the kid must be held absolutely still—kids do not like this and will shout as soon as held down. The holder must not be put off by this, but must hold the kid firmly on his knee, upon a sack for preference, with its head outstretched towards the operator and its tail firmly tucked in under his chest as he leans forward to hold its head.

The first application is with the head of the iron flat down upon the horn bud, the central hole covering the circle so that the bud is clearly seen to stand up above the skin when the iron is removed after the operator has counted eight seconds. He will use no pressure, but hold the iron in place, whilst the holder blows away the smoke caused by burnt hair.

The second application to the same horn bud is with the iron held side down, rubbing gently until the horn bud is flat. It is usual to make the initial application to each horn bud first, then the second step is again to each bud and, finally, feeling the head to ensure that there remain no notches which will grow into unsightly 'slugs' later on, replacing the kid upon the floor and removing the irons from the fire.

Do not over-heat the kid's head by following one application too closely upon another; experience will show that by the time a used iron is re-heated to the essential cherry-red condition, the kid's head will have cooled down.

The disbudded kid will need no further attention, and the hair will grow over the disbudded area, so that by the time the kid is a few weeks old, there will be nothing to show.

The only precaution is to make sure that, if the kid is still with its dam, she does not persistently lick the top of its head. This, in fact, is very seldom the case, but has been known to make a kid's head sore.

The Caustic Method (not recommended)

This chemical disbudding method with a caustic stick was in use before the first disbudding iron reached us from America. In the Author's opinion this was extremely cruel and very hazardous, as it was not 100% successful and the kid cried for hours after the operation. The caustic could run down the head and blind or disfigure the kid which made it altogether a nasty, messy and painful way of disbudding.

HOOF TRIMMING

The feet of the goat consist of an outer wall, an inner sole and heel. The outer wall would, in nature, be worn down by the craggy environment of the mountain types, which have much harder hooves than the desert types. The latter would probably keep in fairly good shape with sand friction.

Diagram 17-2—Overgrown foot

In more civilised conditions, where the goats are pastured on soft turf and hardly ever walk on any really hard ground, the wall grows over the sole (unless it is trimmed regularly), and this can trap dirt and stones, giving rise to lameness as well as eventually ruining the shape of the goat's foot.

Requirements for hoof-trimming are as follows: a sharp 'lamb foot' or shepherd's knife; a Surform file and a pair of foot-rot shears.

Method

Tie up the goat or, if a young kid, have someone hold it. Pick up a front foot and, with the shears, carefully trim back the wall until it is level with the sole. Now trim the heel until it is flat and level with the sole, using the knife and cutting only the very finest slivers off at one cut. Always watch carefully that no further cutting is done once there is the faintest sign of the pink 'quick', either on sole or heel. If you can see this, you have cut too much at one cut, and since it is particularly difficult to gain access when trimming goats with black feet all the more caution must be used.

Continue with the back foot nearest you, holding it firmly in your knees, as in the diagram. Then move the goat over, so that the other side is against the wall, and continue with the other front and back feet. A final filing with the Surform file is beneficial. When you have finished, lead her on to a level surface and make sure that her stance is level and correct. **Trim feet every month.**

Badly Trimmed or Neglected Feet

If you have taken on an animal with neglected feet it will take many regular monthly trimmings to get her right. It may be best to attend to her more frequently at first, perhaps fortnightly. Her feet will have to be reshaped, and this can only be done in stages, taking a little at one time until the foot is the correct shape again. Horseguard Hoof Oil applied daily will help to restore normal horn growth and disinfect any holes or cracks.

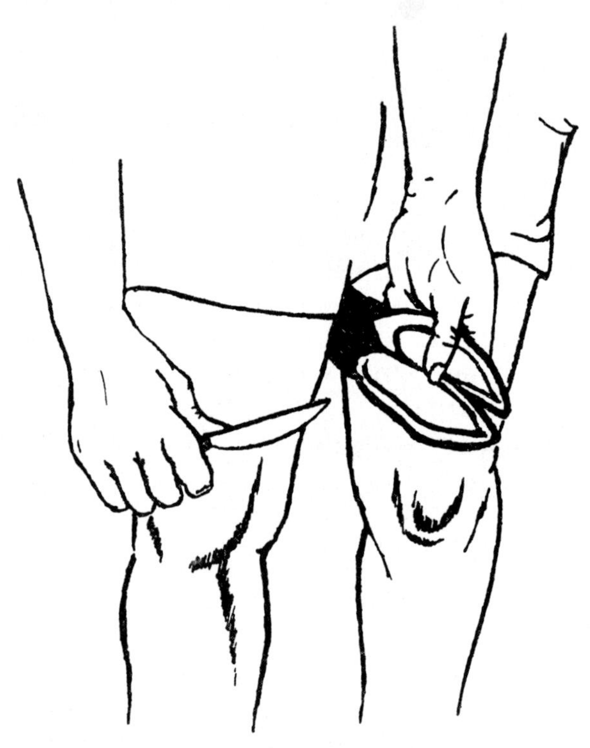

Diagram 17-3—Holding the Hind Foot with the Knees

GROOMING

In the spring the goats will probably be prepared for the coming show season. The winter coat will be coming out and the new finer coat will be growing. Whilst no amount of grooming will make an unhealthy goat look good, an animal in good condition will soon begin to improve simply by being brushed daily, even though for a few minutes only.

Equipment required is a small dandy brush, a body brush, a rubber curry and an old silk scarf or piece of velvet. Use the dandy brush for the long hair, the body brush for the smooth and the rubber curry not only to clean the brushes, but also to bring out the loose, old coat. Finish with the soft cloth.

ROUTINE WORMING

Three weeks after kidding, and again in the autumn, goats should be dosed for worms. They may need another dose between these routine treatments, but much depends upon pasture management. The two drugs preferred at the present time are Thibenzole and Nilverm. Some goats react adversely to the former, whilst others do not seem to notice it. In any case, it is usually only a temporary upset with a short-term drop in yield.

The liquid Thibenzole is easier to use than the tablets. An old plastic shampoo bottle is ideal for **drenching** providing it has a narrow neck and it cannot be chewed or broken.

Dosage

A kid about three months old will require $\frac{1}{4}$–$\frac{1}{2}$ oz, and an adult from $\frac{3}{4}$ to 1 oz, according to size and breed.

Liquid Nilverm is better than the injection and the doses for this are $\frac{3}{4}$ oz for an adult and $\frac{1}{2}$ oz for a six-month old kid.

Although not a routine dose, Nilzam should be mentioned for cases of lung worm. This may be identified in the faeces sample and must not be neglected; so also may Fluke, a parasite for which a snail is the host, and one which is not common to all grazing but is usually found in boggy land. Fluke can kill quite quickly, being particularly dangerous to kids.

Drenching

Using the plastic bottle, hold the goat's head with your left arm, but do not tilt it at too great an angle. Insert the bottle neck into the gap in the side of her mouth and pour slowly into the corner, holding her head to you; if she fails to swallow, gently rub her throat and make sure that she is not holding her tongue in the top of the bottle! Should she cough stop at once and wait.

TO MAKE A GOAT LEAD

Home-made leads are not only cheaper but they also have the advantage of fitting exactly the animal concerned and, with the adjustable slip over the goat's head, it becomes tight enough to prevent it coming off without becoming so tight that the animal is choking. They can also be washed and, if made from nylon cord, can be boiled time after time for show whiteness.

Nylon cord comes in several thicknesses; fine for kids, thicker for adults, and very strong for males. It can also be bought in various colours.

Binder twine, which also comes in coloured nylon now, can be used for home use, but of course the inevitable knots do not help. For an average length lead, you will require 12 yards of nylon. Do not make the lead any shorter or you will not be able to walk on and expect the animal to follow—so essential in lead-training.

Method

Take the 12 yards and divide into six equal strands. Fold the middle of the strands over a door latch or small knob and start to plait with the six strands (three-*twos*).

Continue thus for about 8 in, according to the size of the goat's neck. Then divide to two plaits of three strands each for the slip part of the lead. This must be long enough to allow the goat's neck through the loop which you are now making. At least another 8 in of two three-strand plaits will be needed.

Now put the original loops from the knob or latch, over which you started your first plait, on to one of the two three-strand plaits. This will slide up and down, making the adjustable slip.

Diagram 17-4—A Home-made Goat Lead

Diagram 17-5—Simple Goat Rug (sideways)

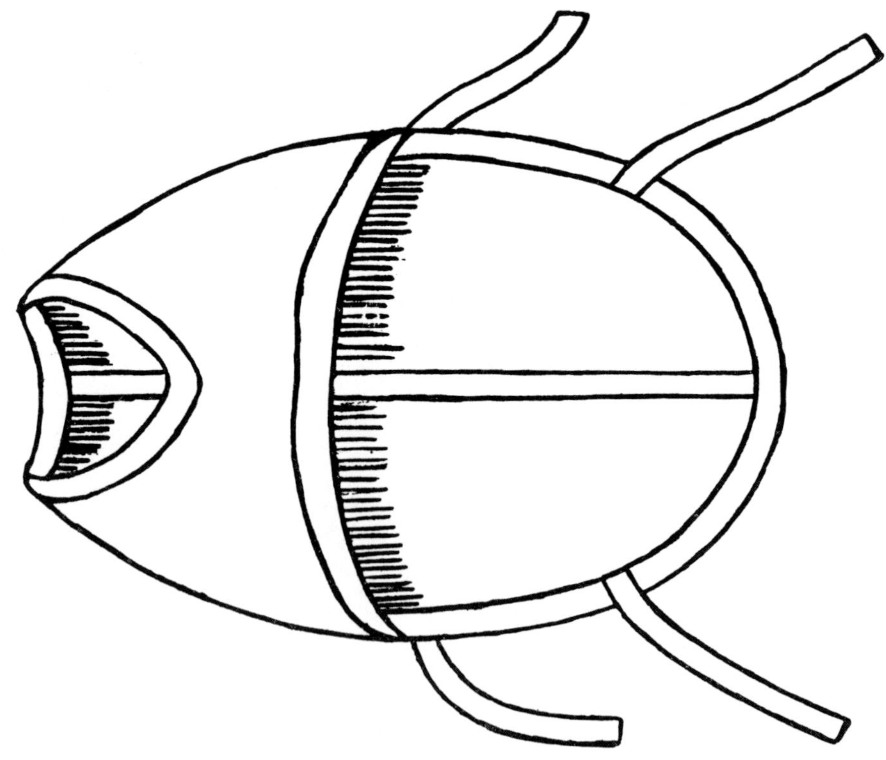

Diagram 17-6—Plan for simple Goat Rug (top view)

Continue down the lead towards the end with one plait of all six strands again, until you reach the end of your six strands. Now either make a neat knot or finish with a professional bind.

MAKING A GOAT RUG

A well-fitting rug which will not come adrift and tangle round the goat's feet is not only useful in times of sickness, but to keep the animal clean and warm at shows, overnight and also after kidding, for the first few days, if the weather is very bad.

Material

A closely woven sack will do, if lined with an old blanket, any type will suffice, but part of an old horse blanket is ideal. A discarded electric blanket may be used, with all the wires pulled out, or an old tweed coat. Almost any material which is warm and not slippery will make a good goat rug.

Method

First measure your goat, from neck to tail and from withers to the top of front leg; also take a note of her girth measurement for ties.

The slip-over pattern is best, with a hole large enough to pass the head through and no tying at the chest. The fastening at the rear end may be by means of leg straps like a New Zealand horse rug, or by means of a tie just in front of the udder.

The outside edges should be bound for strength and neatness, and a length of this binding may be used as a girth tie, having been attached at the point behind the withers on the rug, directly above where the front legs will be, when the rug is on the goat.

Group of Healthy Goats

CHAPTER 18

EQUIPMENT AND MILKING

CLEANING AND STERILISING

Whilst the commercial herd owner will clearly require large quantities of dairy detergent and steriliser, the household goatkeeper who has perhaps one large lidded bucket and one or two small milking buckets, with a small size strainer to complete his milking equipment, can well manage to keep up a satisfactory dairy hygiene by simply steam sterilising these few items.

The strainer may be placed inside the large lidded bucket, together with small cans used for putting into the refrigerator daily, and the large bucket then put over a very slow heat with a small amount of boiling water inside, so that the bucket and its contents are steam-sterilised for 5 minutes. Each milking bucket may then have the same attention.

All items used for milk and milking are always rinsed in cold water until clear, before washing in hot water and detergent. They should be sterilised once daily and turned upside down to keep free from dust.

OFF-FLAVOURS IN MILK

There are various taints which can occur in any milk. Some are through failure to observe the simple sterilisation rules given above, whereas others are caused through storing the milk in open containers next to strong-smelling foodstuffs like onions, oranges, etc. A third and probably the most common reason is an unhealthy state of the udder of the goat herself.

MASTITIS

There are many kinds of mastitis—some giving rise to a horribly metallic flavour in the milk, some making it bitter, some causing discolouration. When in doubt, have a sample

Diagram 18-1—Top: *Vegetable Strainer which makes a good cream skimmer* Bottom: *Vegetable Mill in use as a milk strainer*

of the milk of each goat tested by the County Laboratory. The cost is about 80p per sample.

Milk which tastes or smells of swedes, kale and other feedstuffs, is simply milked too near to the feeding time. Always feed kale and roots *immediately after* milking and never about two hours before milking.

ART OF MILKING

To buy a goat in milk, before one can milk her, is the height of folly, yet so often this mistake is made.

A large commercial herd of goats is usually milked by machine—there is a business which manufactures an adapted machine suitable for this purpose. The exhibitor with the showing of his goats as his hobby is unlikely to use a machine because the goats will be milked by hand at shows and a sudden change from machine to hand is not advisable. Nor will the householder with two or three goats find it worthwhile as goats are normally very easy and quick to hand-milk.

If possible, learn to milk on a goat which has to be dried off prior to kidding, not on one which is in full milk. Another possible way to learn to milk without loss of flow, is to use a maiden milker for a start. This is a goatling which has come into milk before being mated, a phenomenon much more common in goats than in cows.

Since a large yield is not required and normally these animals are not milked right out, or stripped out, no harm will be done providing that extreme gentleness is the keyword for the immature animal's udder.

It is usual to milk the maiden milker once daily only, or whenever it is apparent that the udder is becoming tight, which must not be allowed. The idea is not to obtain as much milk as possible, but to lessen the risk of mastitis through overstocking.

Washing Udders

Opinions differ as to the desirability of washing the goat's udder before milking. Much depends upon the cleanliness of her living quarters. If she is taken from a clean dry pen and milked upon a bench kept solely for the purpose, a wipe round with a dry cloth to remove any seeds or wisps is usually sufficient. On the other hand, if her bedding is not clean and dry, or she has travelled home through muddy gateways, she will require a cloth wrung out in a mild non-irritant disinfectant, one cloth being kept for each goat.

The writer believes that washing twice daily should be avoided if possible, because the natural protection of the skin may be removed and a condition set up whereby various udder complaints are more likely to occur. Certainly the milker's hands must be scrubbed and should be dipped between each goat's milking, if there is the slightest suspicion of an udder complaint.

Letdown

Letting down her milk is not a compulsive action, like blinking, and the goat is able to hold it up, let down just so much, or let down all she can, as she pleases. It therefore behoves the milker to encourage 'letdown' and, having achieved this, to milk out the animal with all possible speed, for the period of *letdown* lasts only a few minutes.

Contributory causes of *letdown* are the wiping of the udder, rubbing the hands around the udder, the clanging of milking buckets and the swishing sound of the next goat's milk going into the bucket.

It will be realised that one does not want to promote this *letdown* until one is ready to start milking, so that everything required should be ready.

Requirements

The essential requirements are a 1 gal size stainless steel or heavily tinned milking bucket, a collecting churn or large bucket with lid (to hold all the milk) and a strainer with filter pad in place over the larger pail or milk churn.

The goat's concentrate ration should be made available if she is to feed whilst being milked. Some goats will not eat until milked, while others prefer to eat whilst being milked and some eat best afterwards. It does not greatly matter which course is chosen so long as it becomes routine. Likewise hours of milking should be adhered to, a 12-hour interval being ideal.

A short-legged milking stool may be home-made, or the legs of a cow milking stool may be cut down to size. If a milking bench with stanchion is used the milker will sit upon the edge of this.

Method

A teat is held in each hand, taking care not to spread the hand on to the udder above the teats. The fingers are closed in turn to expel the milk from the teat, **without any arm movement.** As each teat is emptied, the fingers release, allowing the teat to refill. There must be no pulling down or stretching of the teat and no jerky arm movement. The fingers should work quickly and gently expelling the milk by pressure against the palm only.

When all the milk which can be obtained readily is milked out, the hands are run round the back of the udder, which is gently kneaded to encourage the last drop to flow. Stripping out is done with the fingers as before, not by stretching of the teat downwards with finger and thumb.

The milk is at once poured through the strainer into the large churn or pail and, if possible, the churn stood in a trough of cold water until each goat's milk is added and cooled, when it can be removed to the kitchen refrigerator or dairy.

Stall fastening.

English Milch Goat

CHAPTER 19

PRODUCE FROM THE DAIRY

SCALDED, OR DEVON CLOTTED CREAM

Method Without Using a Cream Separator

Each day's milk is strained and cooled and poured straight into a large flat pan or basin, holding 1 or 2 gal. The pan is set in a cool place or in the refrigerator (taking care not to freeze). After 12 hours the pan is placed over a very slow heat, if necessary using an asbestos mat to slow the heat down.

When the milk crinkles and cracks on top remove the pan from the heat; do not let it boil. Place over cold water until cool enough to return to the bottom shelf of the refrigerator, or the dairy shelf, and leave undisturbed for another 12 hours.

Using a saucer, or skimmer, remove all the cream in one unbroken 'blanket'.

Method Using a Separator

Separate the cream with the thickness adjustment screw at thick or double cream, then gently pour back on to a little skim milk in a shallow cream pan. Set this pan over a slow heat until the cream wrinkles. Then cool and re-skim in one blanket, as above, with saucer or skimmer.

BUTTER MAKING

It is not essential to have a cream separator for making butter, although if more than 1 or 2 gal of milk are to be separated at one time, it will be quicker and more thorough.

Butter From Clotted Cream

Using the method describing the making of Devon, or clotted cream, collect up several days' produce and store in a small glass butter churn or in a large mixing bowl. During the summer months, three days' cream might be collected, and during the winter, one week's cream. On churning day, add no more cream, but stand the churn or bowl at room temperature, or warm over water to a temperature of 57°F ready for churning.

Have all utensils well scalded, and the wooden pats and board rinsed with cold salt water after scalding.

If in a small glass churn, the cream will take only a few turns to show grains, so be careful not to churn into a greasy lump but stop and add very cold breaking water at the rate of one quarter of the amount of cream in the churn. Turn again until the grains can be seen to separate from the butter milk.

Draining and Washing the Butter

Strain off the butter milk, using a piece of butter muslin across the top of the glass churn. Add cold water, turn the handle three times each way, slowly, then strain off the water.

Now tip the butter grains out into a colander lined with butter muslin and wash under a gentle flow of cold water until the water runs clear. Then leave in brine for 20 min. The brine is made with 1 lb salt to 1 gal of cold water.

Patting Up

Using the clean butter pats and board, pat and work the butter, until no more water is exuded, and then work up into pats or blocks of the desired size. In very hot weather it may be best to leave the butter in a cool place, or in the refrigerator for an hour before the final working up.

Keep the buttermilk for scone and cake making, as it is a natural raising agent and makes beautiful tea-time luxuries.

Butter with Separated Cream

Add each day's cream to the glass churn, as for the scalded cream method, adding no more 12 hours before churning; then mix well. As before, bring up the temperature of the cream to 57°F before beginning to churn. Churn steadily until the cream becomes very thick and heavy to churn, then stop and add breaking water. Churn again until the butter breaks and proceed as above from that stage onwards.

The yield of butter from the milk will be more from the separated cream, but the residue of skim milk from the scalded cream will be of better value.

YOGHURT MADE IN A SMALL WAY FOR THE HOME ONLY

The requirements for yoghurt are as follows:

1. A wide-mouthed thermos flask.
2. A double saucepan.
3. Yoghurt culture or prepared yoghurt.
4. Two pints of fresh goat's milk or skim from making scalded (or Devon) clotted cream.

Method

Boil the milk in the double saucepan and hold it at simmering point for 30 min.

Cool the milk by circulating cold water round the saucepan until it is at 113°F. Pour the culture or the prepared yoghurt into the clean flask, pour on the milk at exactly 113°F as quickly as possible and cork the thermos flask.

Leave the flask for 12 hours, after which pour out the curd, leaving the flask unwashed so that the remnants will inoculate the next lot of yoghurt to be made.

When the culture becomes tired and loses its acidity, or becomes gassy or wheyed off, replace it with new culture or newly prepared yoghurt.

Toggenburg milk having good solids-not-fat, and moderate butterfat, makes a good sample of yoghurt, but for slimmers the low fat yoghurt made from skim milk will be preferred. Those wishing to make yoghurt in a big way for sale should study the methods advised in specialist publications.

CHEESEMAKING*

Making cheese from the milk of the goat is one of the most profitable ways of using milk which may in spring and summer be surplus to liquid milk requirements, particularly as a well-made cheese will keep for a long period.

* Contributed by Anne May, a specialist in the art of cheesemaking.

Many of the most delicious European cheeses are made with goats' milk, and some with ewe and goats' milk, mixed.

Surplus milk may be stored in a freezer until several gallons have been collected, when it may be defrosted to make a hard cheese, such as a smallholder type. Even without the freezer, goat milk will keep for several days in an ordinary refrigerator.

A hard cheese may be made with as little as 2 gal of milk, but a 2 lb cheese will tend to dry out rather more quickly during the maturing process.

Almost any of the cheeses normally made with cows' milk can be made with goat milk, but some require more skill than others. It is suggested that a start should be made with one of the simpler ones, then some experimentation will show which types are more suited to the premises and also the individual tastes.

Quality

Milk of the highest quality is required for making cheese. Any off-flavours in the milk will taste even worse in the eventual cheese. The bacteria causing these off-flavours may also spoil the texture of the cheese and prevent it keeping as well as it should. Some cheeses are better made with heat-treated milk and others are not.

Small-scale Cheesemaking

When making cheeses for the family only, it is an advantage to have a warm kitchen to work in, as warmth is essential during the early stages. Do not handle fruit, or make bread, at the same time as cheesemaking, as yeast contamination from these can spoil the cheese.

Time

Until the recipe has been read and fully understood, it is essential to concentrate upon it, but once it becomes routine, other household tasks may be fitted in and cheesemaking is not so time consuming.

Cleanliness

All work surfaces must be clean and free from clutter. A clean overall and hair band are assets. Have by the table a pan of boiling water in which to dip used cloths and utensils.

Requirements

Diagram 19-1 shows most of the basic requirements for simple cheesemaking, many of which would be found in an average household. *Diagram 19-2* suggests the sort of bowls, racks and other implements and utensils used to collect whey, and in which to work the cheese. Most households have a preserving pan, usually holding about two gallons and having a wide surface area.

True jacketed vats are almost unobtainable and it is not necessary to buy a lot of specialised equipment; however, it must be remembered that the cheesemaking qualities of milk are spoiled by overheating, so do use gentle heat and an asbestos mat or heat diffusing plate under the container and stir frequently until the correct temperature you require is reached. A thermometer is essential, but it need not be the expensive floating dairy one, although these are excellent and have very clear markings. A cheaper thermometer can usually be obtained either from a chemist or a chemist's supplier, and as long as it shows a range from 0°F to 170°F it will suffice. Renneting usually takes place at between 85°F and 90°F; 'starter culture' is added to the milk at from 90°F and 100°F; and for heat treating the milk, when this is done, the temperature is raised to 150°F.

SOME NOTES ON THE USE OF 'STARTER CULTURE'

The term 'starter', when used in cheesemaking, refers to a concentrated culture of the correct bacteria to help to develop acidity in the milk during the early stages and to assist the development of flavour and aroma in the finished cheese. These bacteria are usually present in fresh milk but they have to share this medium with a host of other bacteria, not all of which are harmful but all of which multiply rapidly at varying temperatures. When the temperature of the milk is raised for pasteurising or heat treating most of the bacteria are killed. If a culture of fresh lactic bacteria is then added to the milk at the required temperature, these bacteria are free to multiply without competition from others. In simple terms, this is what is meant by the term 'starter culture'.

It is possible to incubate lactic culture from a sample of clean fresh milk straight from the goat, but it is better to make use of the excellent and refined services of the

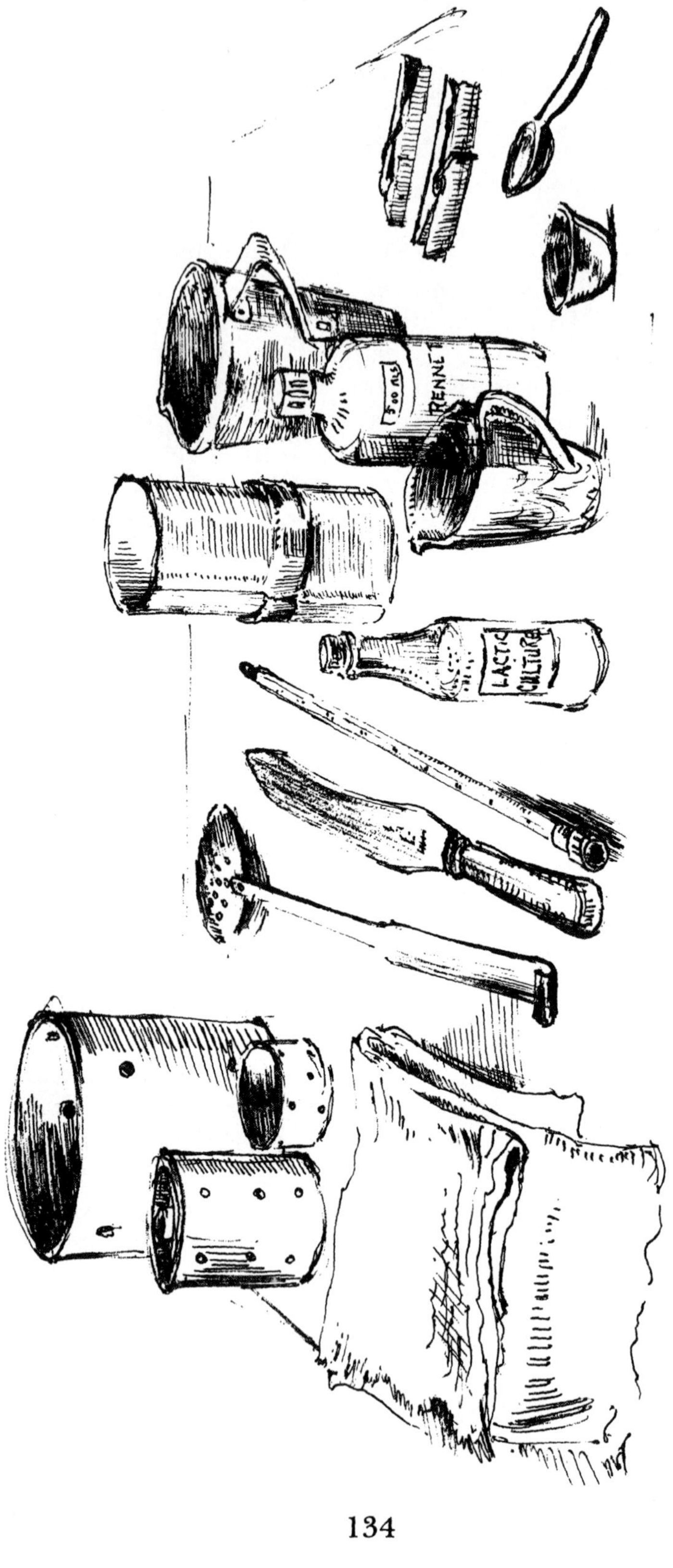

Diagram 19-1—Equipment Needed for Making Cheeses at Home

laboratories. Lactic culture can usually be obtained either in liquid or freeze dried sachets from the stockists mentioned in the Appendix at the end of this chapter. This culture can be made to last for several months if care is taken. Prepare a screw-topped jam jar and lid by boiling in water for several minutes, fill seven-eighths full with milk and boil this by standing in water and raising the temperature gradually to boiling. When the milk has cooled to below 90°F add a small amount of the freeze-dried powder or the liquid culture and half screw on the lid of the jar. This transfer of culture may be done over a flame for extra cleanliness. Bring the temperature of the inoculated milk back to 90°F and incubate by standing near an all-night burning boiler, or in an airing cupboard with unlagged tank that is continuously hot, or even by placing the jar with its lid tightly screwed in a wide necked thermos flask filled with water at 100°F. If using the latter method the water may need reheating after several hours. Incubation should be complete within about twelve hours, if the culture is fresh and active, when the milk will have become thick but not separate and should smell cleanly acid and faintly 'cheesy'. Take another jar of boiled and cooled milk and add some of the new starter to this. Keep this jar of inoculated milk, duly labelled, in the refrigerator for up to a week, ready to incubate when required, whilst using the first jar for cheesemaking immediately.

SOFT CHEESES

In soft cheesemaking the milk may be coagulated by using lactic starter or rennet and the curd is usually drained through muslin or cheesecloth, or put in a mould, with both curd and whey being ladled into the mould and the whey draining through the curd. The curd formed by coagulating goat's milk is much softer than that formed from cow's milk and it is sometimes impossible to form a curd that is sufficiently firm to cut without the use of rennet, but this need not affect the end result. In order that the whey should drain freely and to avoid the curd being chilled, a warm atmosphere is essential, and the temperature of the room in which you are working should not fall much below 70°F.

Soft cheeses are usually eaten within a few days of making and have a limited keeping quality unless kept in a refrigerator or freezer. Exceptions to this generalisation are

Diagram 19-2—Equipment Necessary for Home Cheesemaking

the *Colwick* which, when made with untreated goat's milk, can be ripened to a full flavoured cheese with a soft consistency, and the *Pont L'Evêque* which has the whey removed from the curd fairly rapidly before being moulded, and which is also matured.

LACTIC CHEESE

Since cleanly produced goat's milk sours very slowly, and if left to do so naturally may pick up a lot of unwanted organisms on the way, a very simple way of coagulating the milk is by the addition of commercially produced lactic acid. Although this recipe is simple and produces a very palatable curd, with a variety of uses, it is not considered by nutritionists to be a true 'cheese', and now that lactic acid is difficult to obtain in small quantities and expensive when you can get it, this means of forming a curd is less frequently used than formerly.

Method

Heat fresh milk to 110°F and remove from heat. Add 1¼ teaspoons lactic acid per pint of milk used and stir well until the milk coagulates and becomes granular. Leave to stand for an hour. Tip into a muslin or cheesecloth over a bowl or bucket. Tie up and leave to drain until firm. The whey should be clear and yellow. When sufficiently drained remove to a dish and add salt to taste. If desired, herbs may be added.

When draining soft cheeses through a cloth, difficulties may sometimes be experienced if quantities larger than a gallon are drained in one bag. Larger quantities of curd drain much better if the cloth is draped over a bucket, to catch the whey, and attached to the top of it with clothes pegs, then covered with a lid or part of the cloth, instead of being tied in a bag. When using this method of draining it is also much easier to scrape the curd down than the messy method of transferring the curd from one cloth to another.

CROWDIE

This cheese has a less granular texture than lactic cheese and a more interesting 'cheesy' flavour.

Ingredients and equipment

1 gal whole or skimmed milk.
2 lbs. 'starter' (yoghurt can be used).
Salt.
Large pan; i.e. preserving pan.
Thermometer.
Ladle.
Coarse textured cheese cloth.
Bucket or container for whey.

Method

Heat-treat milk, if desired, by heating to 155°F then cooling to 90°F. Stir in starter and leave, covered, in a warm place overnight for acidity to develop and a curd to form. When thickened, place pan on low flame and raise temperature to 100°F whilst slicing the curd gently with the ladle. Maintain this temperature for half an hour, stirring from time to time. Leave for 10 min to settle. Scald the cloth and lay across bucket, secured with clothes pegs, and ladle in the curd with scalded jug. Leave in a warm place for several hours, scraping down the curd when necessary until all the whey has drained off. Stir in salt to taste and either shape into rolls or squares or pack into small containers with lids. If possible, leave no air space in packing as undesirable moulds can grow on the surface of soft cheeses.

COLWICK

This cheese is simple to make, does not take a great deal of milk and is extremely versatile. It can be eaten fresh as traditionally it used to be when made with cow's milk, when it has a clean acid flavour. This lends itself extremely well to the dish-shaped cheese being filled with well-drained fruit and topped with cream. Try it for a special after-dinner dessert. Alternatively, it can be wrapped in cling film when drainage is complete, and ripened.

Ingredients and equipment

6 pints of whole milk for one cheese.
1 tsp. cheese rennet diluted with 6 tsp. water.
Salt to taste.
Preserving pan or other large pan.
Thermometer.
Ladle or slice.
1 round mould; e.g. Boddington's 2½ lb Stilton mould.
Draining mat.
Cheesecloth.
Board or rack.
Container to catch whey.

Method

Do not heat-treat the milk if you wish to mature this cheese. Heat milk to between 85°F and 90°F and stir in rennet. Do not overstir. Cover and leave to coagulate for 50 min or until firm and no milk adheres to finger placed gently on surface. Place the board or tray on the container for catching the whey with the draining mat on top. On top of this place the mould which has been lined with the cheesecloth so that the cloth hangs over the top of the mould. Ladle the curd into the mould in thin slices, gently

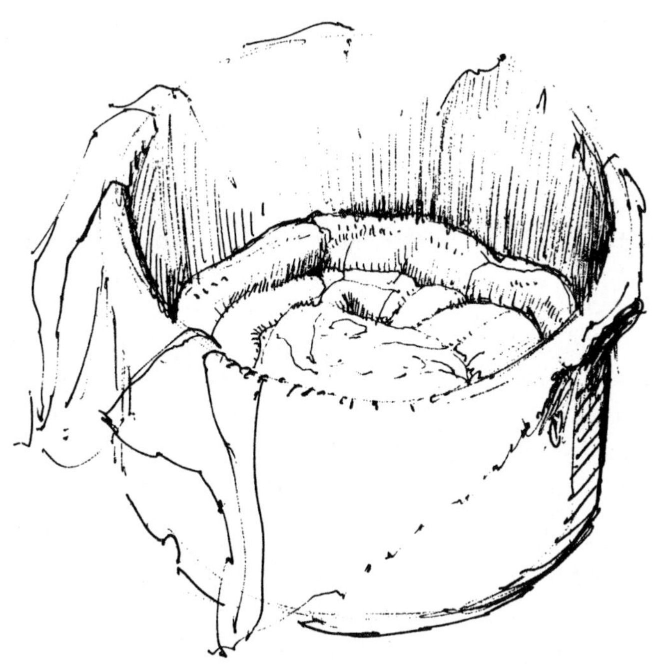

Diagram 19-3—Colwick Cheese—Curd Draining in Mould Showing Curled Edges

drawing the cloth up as you fill the mould. As the curd sinks in the mould, which it does most quickly during the first hour, pull the muslin or cloth upwards in order to make the edges curl over and inwards—this helps to produce the dished top which is characteristic of this cheese. If you wish you can tie a 'Stilton' knot, by taking three corners of the cloth and winding the remaining corner round these three. Leave the cheese to continue draining and after three or four days it will be firm enough to handle. If salting is desired rub a dessertspoonful of salt through the cloth before gently removing. If the cheese is to be matured, wrap it firmly in cling film and leave in a warm room, at about 60°F for two or three weeks. The outside of the cheese may turn a pale shade of beige and the inside will become soft. Should the cheese begin to smell remove to a cooler room and the smell will disappear, but it is better to start off in a warm room and the curd should never be allowed to become chilled.

PONT L'EVEQUE

This delicious small cheese is great tun to make and will bring a truly 'Continental' flavour to your dinner table. It develops an interesting, firm but edible crust, whilst remaining soft inside. Traditionally made in Normandy and moulded in an oblong mould, it can be moulded in the smaller ring of the metal *Coulommier* mould or in the ½ lb plastic Boddington's mould. As this cheese should not be allowed to develop any acidity in the making process it is better not to add 'starter', and the milk is not heat-treated, but should be absolutely fresh.

Ingredients and Equipment

1 gal fresh whole milk.
1 tsp. rennet diluted with 6 tsp. water.
Salt—1 level tsp. per cheese.
Large pan, preferably with lid.
Ladle.
Carving knife.
Thermometer.
Cheesecloth.
Draining mats.
Draining racks—cake cooling trays or well-scrubbed wooden seed trays can be used.
4 small moulds, 3½ in diameter.

Method

Warm milk gently to 90°F and stir in rennet. Cover with lid or cloth to conserve heat and leave for 45 min until curd is set firm. Scald the knife and cut the curd into squares and then diagonally across. Leave for 5 min. Meanwhile set up the draining equipment by placing the draining tray across a bowl or trough to catch the whey and lay the cheesecloth across the tray. Now scald the ladle and ladle out the curd in thin slices into the cloth, where it will drain rapidly. When all the curd is in the cloth fold over the edges and cover with a clean tea-towel wrung out in hot water, replacing the hot cloth as drainage continues, in order to keep the curd warm. After about 30 min, when the curd should still be wet, soft and flaky, cut into squares to facilitate drainage and leave for a further half hour, by which time the curd will probably be firm enough to come away from the cloth cleanly at the edges, but should not be too dry.

Set up the moulds on a draining mat placed on a tray or board over a shallow tray such as a meat roasting tin. Break up the pieces of curd and place them firmly in the moulds. When the moulds are half-full, sprinkle salt on to the curd then fill up to the top. Put another mat on top of the moulds with another rack or board on top and turn the cheeses. Repeat this turning process every quarter of an hour during the first hour and then less frequently, but at least once a day. The cheeses will be ready to have the moulds removed after about three days. Continue turning for a further three or four days then wrap in waxed paper (that from cereal packets is ideal).

Store on a shelf to ripen at a temperature of about 60°F. Turn the cheeses from time to time and, in about three or four weeks, the firm crinkled coat will have developed and the cheeses should feel firm but soft to touch and are ready to eat.

HARD CHEESES

Most hard cheeses have the curd, or 'junket', treated in various ways so that the whey can be removed before the curd is put into the moulds, whether the cheese is then put under pressure or not.

Smallholder

This is perhaps the most popular of the home-produced hard pressed cheeses and, once you have familiarised yourself with the method, one that is surprisingly easy to make. It can be made with as little as 2 gal of milk, although a cheese made with 3 or 4 gal can be matured for longer without drying out. If the cheese does dry out too much, try putting it through the mincer and using it for cooking—it will make delicious sauces and souffles—or grill it on toast for Welsh Rarebit.

Ingredients and Equipment

3 gal. whole milk.
3 tbs. lactic starter.
1 tsp. rennet diluted with 6 tsp. water.
Salt.
Preserving or other large pan.
Thermometer.
Ladle or slice.
Long bladed knife.
Horizontal curd knife (not essential).
Close-weave cheesecloth—for containing curd during 'cheddaring' process.
Muslin—for lining mould (cheesecloth can be used if necessary).
Tray for catching whey from press.
Bucket.
Measuring jug and teaspoons.
Cheese mould; e.g. cake tin or coffee tin with holes pierced in can be used temporarily, or Boddington's 2½ lb Stilton mould with wooden follower is ideal.
Weights; e.g., bricks, or press.

Method

All equipment should be sterilised before use. Starter is optional if you are using untreated milk but necessary if the milk has been heat treated. Heat the milk to 150°F and cool to 90°F. Add starter, stirring well, cover and leave for 30 min for acidity to develop. Add the diluted rennet and stir gently but thoroughly until 'stir' marks appear. Leave covered for 40–45 min, until the curd is firm enough to break cleanly over an inserted finger. Using a knife with sufficiently long blade to reach the bottom of the pan, insert the knife and draw it towards you in a straight line. Repeat this cutting to form ½ in squares. If a curd knife is available, insert the blade down one of the cuts and turn it round the whole way, then lower it slightly and repeat the circular cutting movement until the cylinders of curd have been cut into cubes. Alternatively, with the long bladed knife, cut diagonally through the curd, first in one direction then at right angles. The action of cutting the curd releases whey.

Leave to settle for a few minutes, while washing the hand and forearm thoroughly, then gently loosen the curd by hand from the edges and bottom of the pan.

The curd must now be scalded. Heat the curd very slowly to 100°F, taking half an hour to do this and preferably moving the curd very gently by hand during this time. It is important not to handle the curd roughly at this stage or a lot of goodness will be lost from the finished cheese. The action of heating the curd will cause the particles to contract and expel whey and they will become firmer and much smaller. Allow the curd to 'pitch' and settle in the pan for 30 min away from the heat when the 100°F has been reached.

The whey must now be removed from the curd. To do this, one way is to lay a large cheesecloth over a container—a bucket will do—and tip the curd into this. Remove the cloth bundle of curd on to a baking tray and form into a loose bundle holding the three corners of the cloth and winding the fourth round—this is known as a 'Stilton' knot. Leave for 20 min then pour off the whey again and untie the bundle. The curd will have consolidated into a solid mass and may now be cut into four strips. Re-tie bundle and leave for a further 20 min. Pour off whey again and untie bundle, this time cutting each strip into two, so there are eight strips. Re-tie and leave for 10 min. Untie bundle and carefully separate the strips, piling one on top of another in pairs, turning each piece so that the warmer edge is on the outside. This process is part of the Cheddaring process and is known as 'piling the curd'. Re-tie the bundle and repeat this rearranging of the curd once or twice until very little whey is escaping and the curd is quite firm.

Break up the pieces of curd by hand, into a clean scalded bowl and scatter 1 oz salt over the pieces. Mix well in.

Line the cheese mould with muslin and put in the curd, pressing it in firmly. When the mould is full, pull up the cloth and fold over the top, then put on the wooden follower and put to press with 14–28 lb pressure. Leave overnight to consolidate the curd. Diagram 19-4 shows a way in which a Smallholder cheese can be pressed without using a proper cheese press, and making use of objects found in most homes instead of weights.

Next day tip the cheese out of the mould and remove the follower and cloth. Wring out the cloth in hot water. Reverse the cheese in its cloth and put back in mould, with wooden

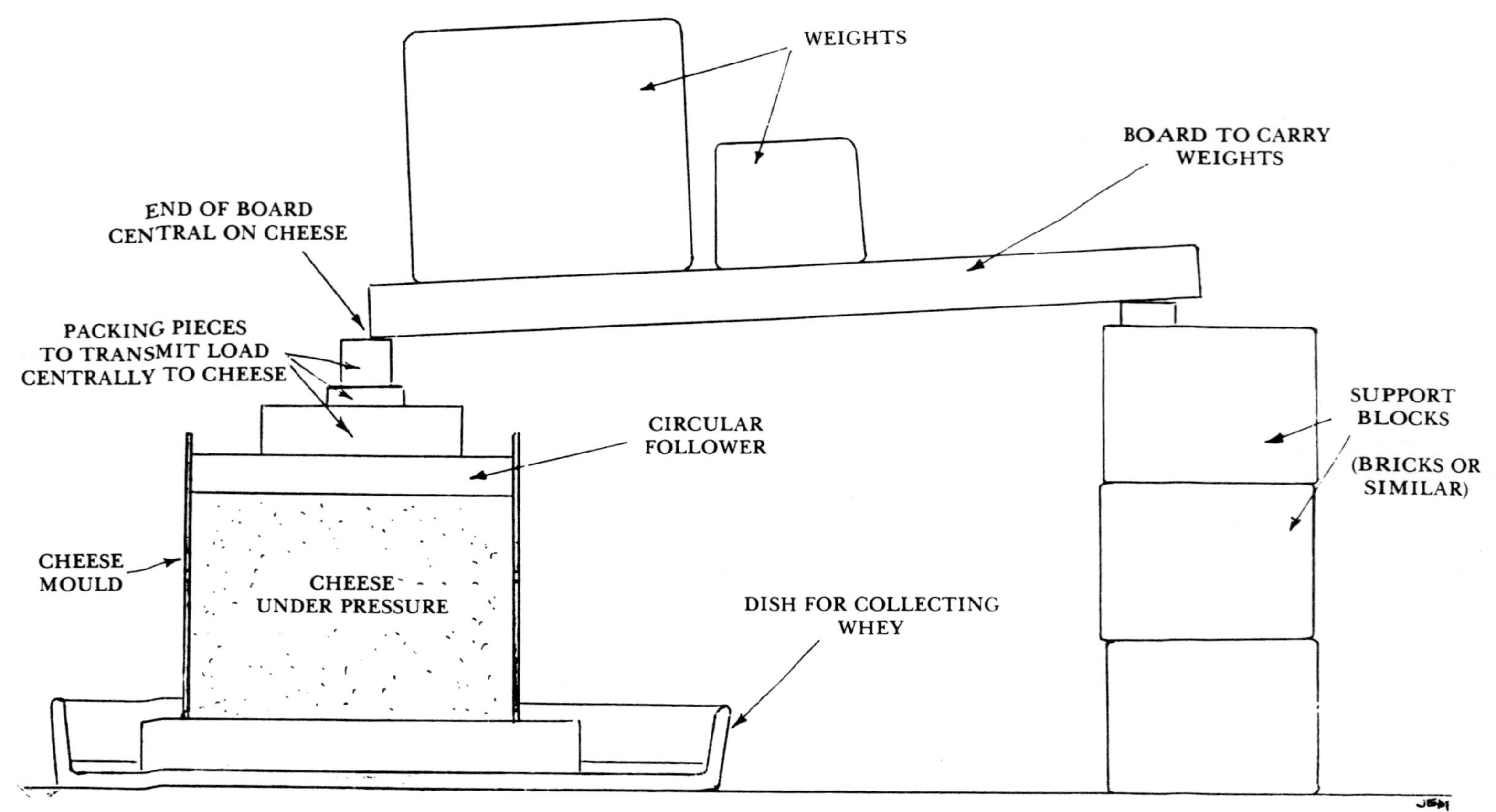

Diagram 19-4—A Simple Cheese Press

follower. Put to press again with more pressure—40–56 lb according to the size of the cheese.

Repeat this procedure each day—removing from mould, wringing out cloth, reversing cheese and rewrapping and remoulding and returning to press for three or four days until the cheese is really firm and no more whey is coming out under pressure. The time taken to reach this stage varies slightly and may take longer to reach in winter. In the summer it is advisable not to leave the cheese under pressure for longer than necessary as the cheese may lose some of its goodness and become dry.

If the cheese is to be matured, bandaging does help to keep the cheese in good condition and prevents drying out as well as making the cheese look professional. Diagram 19·5 illustrates this. Cut two circles of cloth ½ in larger than the

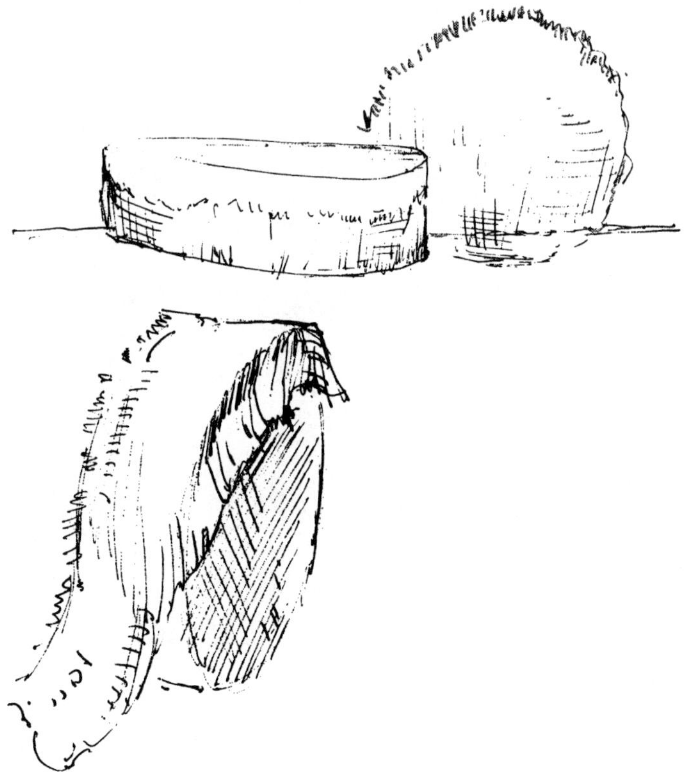

Diagram 19-5—Bandaging a Hard-Pressed Cheese

Photographs of cheeses by Mick Duff
Sketches by Gill Brown

circumference of the cheese and a band ½ in higher than the height of the cheese. Using lard (or edible paste if available) smooth and press on the top circle, leaving no creases and pressing down the sides of the cheese. Invert the cheese and repeat with the other circle. Now put the cheese on its side and smooth on the straight piece of cloth, laying one straight edge level with the top and overlapping the other edge on to the bottom of the cheese. Smooth out any folds or air pockets.

Mature the cheese on a clean, airy shelf at 50–60°F, turning daily at first then less frequently until it is ready. A small 2 or 3 gal cheese will be quite edible in a month but for a more interesting flavour keep it longer, checking that it is not drying out.

Goats in Their Natural Environment

CHANGES IN THE LAW

Modifications have been made to the law since the original edition was published and for convenience these are noted below. Goatkeepers should keep up to date on changes by being a member of the British Goat Society or the Pygmy Goat Club. Also European regulations must be followed where these are applicable in the UK..

1. Goats born on your holding must be ear tatooed or ear tagged before they leave the holding such as when sold, but not for a temporary movement such as visiting a vet.

2. Goats should be registered with the appropriate department at DEFRA, the government ministry that deals with agricultural matters.

3. A Certificate must be obtained from the local Trading Standards Office when it is intended to transport goats to a different area. However, this does not apply for short journeys (under 32 miles) in a vehicle owned by the goatkeeper and the vehicle is not more than 3.1 metres internally.

4. Accommodation, health, feeding, water and other aspects are covered by regulations which are contained in leaflets available from DEFRA in London.

5. When showing goats, regulations will be issued by the organizers and these must be complied with.

If in doubt seek advice from the Trading Standards, DEFRA or the local authority. Obviously, although goats may be permitted in a garden they must not constitute a nuisance such as causing unpleasant smells or noise.

CHAPTER 20

GOATS AND THE LAW

NOTIFIABLE DISEASES

Foot and mouth disease is the most terrifying of those diseases which must be notified by law, but fortunately this dreaded scourge does not often cause the slaughter of goats, which are usually only affected when they are caught up in an infected area, on a farm where bovines have been confirmed with the disease and have been slaughtered. Generally it is thought that, as with tuberculosis, goats have great resistance to foot and mouth disease.

The symptoms are blisters about the mouth and between the claws of the hooves, followed by lameness and fever. Animals do not necessarily die of foot and mouth disease, but become so unthrifty that they are usually unprofitable anyway.

ANTHRAX

Goats can have anthrax, a notifiable disease, but in actual fact it is extremely rare and unlikely to be found in the goats of the British Isles.

MOVEMENT OF ANIMALS

Any person moving cattle, goats, sheep and pigs, is required by law to keep a record of such movements, which must be produced on request by Police or Ministry Officials. Thus if a goat is taken to a show, or taken to stud, this must be recorded, with dates and addresses.

Farm animals are not allowed to stray onto the highway and, if being driven on the hoof to their destination, must be driven by two herdsmen, one in front and one at the rear. A person is required to fence against his own animals, but not to fence out other people's stock.

If goats stray into other people's land and commit damage, the owner is liable, unless he can prove that their escape was caused by an Act of God, or by a third party, unknown to him, or by the plaintiff's own act or default.

If a neighbour allows a poisonous tree to overhang your land, he is responsible if your goats are injured by eating it, but if the leaves do not overhang, then the goats reaching for them and subsequently becoming ill, is not his responsibility.

ANIMAL TRANSPORT AND THE LAW

It would seem that whilst a person may transport a goat with its unweaned kid or kids, one must not transport without a division in the vehicle a goat with another goat's kids. Nor may one take along a male kid of over six months with females. The wording is somewhat obscure and different interpretations could be placed upon the instruction.

Her Majesty's Stationery Office can supply all the necessary leaflets and when in doubt about the legality of an act it is wise to obtain the appropriate information.

THE FOOD AND DRUGS ACT, 1955

Although goat's milk and articles of food derived from it do not come under the Milk and Dairies Regulations; nevertheless, they do come under the jurisdiction of the above act, and it is an offence to offer for sale any commodity which is not of the nature, substance or quantity demanded, or which has been adulterated or diluted.

A reasonably high standard of hygiene may be demanded and premises may be inspected and approved, or disapproved, by the Authorities. There may also be local bye-laws as to the actual keeping of goats on the premises, as well as the selling of produce, and it is as well to find out before starting any such operation, particularly on a commercial scale.

BUYING AND SELLING GOATS

When selling a goat, any member of the British Goat Society is bound, by the very fact of his or her membership, to supply to the new owner a registration transfer if the goat in question is a registered one. Also, the owner or authorised

Figure 11—Toggenburg Buckling

(Bred by the author and exported to South Africa)

agent of a stud goat is expected to give (after service), a certificate of service of the female on the appropriate British Goat Society form to the owner thereof. A form of contract for the sale of a goat may be obtained from the British Goat Society.

If a breeding term agreement is arranged by buyer and seller, to suit their own circumstances, such agreement should be clearly described. Two copies should be made of the agreement—one being held by each party to ensure that all is clearly understood. This may not be legally binding, but in practice it does help to avoid trouble.

CHAPTER 21

ILLNESSES AND POISONOUS PLANTS

WATCHFUL EYE ESSENTIAL

An experienced goatkeeper will keep a watchful eye on the goats, even when the long summer days mean peaceful days spent grazing in the fields and browsing in woodland, for regular check-ups are needed to spot any trouble in the early stages. Early diagnosis is the secret of successful treatment with goat complaints, for if the correct treatment is given at once, a serious illness will be avoided.

No animal responds so quickly as does the goat, but conversely, no animal suffers from diseases so quick to kill, without the speedy assistance of the owner and/or the veterinary surgeon.

THE MEDICINE CHEST

Every form of livestock requires certain items to hand in the medicine chest, and some of these will be readily available in the average household bathroom cabinet.

The goat-keeper who has the following can feel secure against most eventualities:

Clinical thermometer.
Dettol or other non-irritant disinfectant.
Surgical gauze and cotton wool.
1½ in zinc oxide adhesive tape.
Surgical scissors.
Medicinal liquid paraffin.
MacLeans indigestion powder.
Thibenzole and Nilverm.

Measuring glass or jug.
2 in crepe bandage.
Animal Lintex, a packet.
Small plastic drenching bottle.
Epsom salts.
Bicarbonate of soda.
Sulphamezathine, 0·5 gm (50).

A disposable plastic hypodermic syringe with short needles.
3 cc penicillin for injection in emergency.
Standard Bloat Drench
(mix ½ oz turpentine, ¼ oz spirits of ether, and top up to ½ pint with linseed oil).
Tin of Horseguard hoof oil.
Glucose.
Terramycin aerosol spray.

COMMON AILMENTS

Worms

Various types of worm infest the goat, many being common also to sheep, so that territory heavily grazed by sheep is unsuitable pasture for goats. Horses, on the other hand, digest harmlessly those parasites which make goats ill and leave only worms affecting their own kind behind. They are useful, therefore, for grazing off the herbage left by goats, so that the new growth is clean.

Diagnosis

To find out if goats require worming and, if so, which drug to use, send a small droppings sample, in a clearly labelled tin, to your veterinary surgeon, who will then report the count and advise on medicine.

Frequency of Dosing

With well managed pasture, it is usually only necessary to dose about three weeks after kidding and again in the autumn. The method of giving the dose will be found in the chapter on 'Routine work in the Goat House'.

Lung Worm (Husk or Hoose)

Symptom is persistent husky cough; the drug Nilzan is effective, as also for fluke. The fluke is only present in marshy ground where the snail which is its host exists. Extreme loss of condition is usually noticed, but both husk (lung worm) and fluke will be present in the faeces in egg form.

Coccidiosis

Coccidia are very commonly present and a heavy count is particularly dangerous in young kids, although it can also kill adults within a few days if not treated. Whilst other symptoms are much the same as those of a heavy worm count, profuse scouring is not so essentially a symptom. In fact, constipation may be noticed. However, if there is much scouring and bloodstains are noticeable, this disease is then dangerously advanced and veterinary help should be sought without delay.

Sometimes a goat will appear blown with refusal to eat—particularly concentrates—loss of cud and grinding of teeth, a hidebound staring coat and anaemia (showing in eye membranes, and the skin under the tail) are all indicative of one or other form of internal parasites.

It may be advisable to treat the goat without waiting for the result of a faeces count, and in this case the Author always uses Sulphamezathine first.

Dosage

The 0·5 g tablets are used in preference to the 5 g, because these are easier to crush and balling guns have been found dangerous in the hands of novices. The amount of the dose is according to the body weight of the goat, but approximately 12 tablets of 0·5 g are needed for an average sized adult as an initial dose, followed by six tablets every twelve hours, for four days.

Methods

The tablets may be crushed between two folds of strong paper, using a bottle or rolling pin. Tie the goat up in a corner and tilt her head towards you, against your chest, opening her mouth with the left hand and throwing the powder from the palm of the right hand straight on to the back of her tongue. It is not easy to push tablets down a goat's throat without being bitten, and they get broken and lost.

Alternatively, pour the powdered tablets into a drenching bottle on top of a little warm milk. Shake well and drench at once, taking great care not to tilt the head back, but inserting the bottle neck into the corner of the mouth.

Never stop the dosing until it is completed, although the goat may show signs of recovery after two or three doses.

Other remedies

Framomycin powder, Whitsin-S-Sol and Zoaquin tablets are also very effective in combating Coccidiosis.

Entero-toxaemia

Probably responsible for more losses amongst goats than any other disease. Difficult to diagnose until it is acute, the novice seldom acts in time, for a goat, which was apparently normal and well the evening before, may be found dead in the morning.

Cause

Clostridium Welchii, type D is usually responsible, and often strikes those goats in the best condition. The organism is one which is present in the intestines and is not noticeable until triggered off by some outside circumstances, such as the rapid growth of, or change on to, lush pasture; or a sudden increase in concentrates or even a journey for an animal not used to travelling.

A very heavy worm count is likely to be the most frequent cause of the sudden increase and violent formation of poison in the intestines. It is therefore vital to attend to worming.

Symptoms, and Immediate Treatment

In the kid, the first symptom may be an apparent inability to balance; the kid may hop, fall over, run backwards or stagger, before the other symptoms are noticeable.

The adult milker will probably show impending illness by a sudden drop in milk yield. She may also walk a few steps backwards and stretch her body persistently (a sign of colic in the early stages). Her appetite is negligible, her droppings turn to liquid scouring and, towards the advanced stages of the disease, are often blood-stained. Severe colic and straining is followed by coma and death if the correct treatment is not given *at once.*

There is a maximum time of about two hours from the onset to the end with this disease. Three c.c. of penicillin, injected intra-muscularly, and a large initial dose of Sulphamezathine (the latter given by a veterinary surgeon intra-veinously, if the animal is in extremis) or orally if the goat has been seen in time. This treatment should prevent the formation of more toxins and neutralises those already present in the intestines. Hence the importance of using these two drugs together.

Before the days of these two life-savers the only hope was to treat the symptoms whilst keeping the pain down with injections of morphia. Nowadays, the veterinary surgeon has anti-spasmodics such as buscafan and pethidine, at his disposal to relieve pain.

Recovery

After such a serious illness, careful nursing is essential, and attention to the smallest details worthwhile. Natural foods should be given, such as hazel twigs with catkins, in winter, ivy without berries, holly without prickles, the very best hay and, if possible, milk and glucose. In summer, feed brambles and branches of elm, etc.

Keep the goat rugged up and well bedded, free from draughts but with plenty of fresh air and encourage her to drink, offering mollassed warm water, oatmeal gruel and occasionally slightly salted warm water. See that her mineral lick brick is within reach.

Prevention

When there has been one case of entero-toxaemia there may be more. Consult your veterinary surgeon about having the herd vaccinated, but have this injection in the brisket and not just before a show, as a large lump may result temporarily.

Mastitis

Common to all mammals, this complaint is usually brought about by bacterial infection. It is undoubtedly passed from one animal to another by the milker's hands and by flies in summer.

Symptoms

A hard and inflamed udder; tainted, discoloured or bloodstained milk. Clots or strings in the milk. Sometimes the milk curdles and sets solidly like colostrum (beastings) when boiled. Lumps in the udder may, or may not, be caused by bacterial infection; a clue to this are the glands under the tail, which may be felt to be enlarged and harder than usual if an infection is present.

Treatment

It is extremely unwise to leave a case of suspected mastitis untreated for it may become chronic and part of the udder cease to be functional.

A sterile bottle should be tilted under a well cleaned teat and the fore-milk carefully drawn direct into it. The bottle is then labelled with the goat's name, the owner's name and address, and sent to the veterinary surgeon for his despatch to the laboratory which will carry out the tests.

The veterinary surgeon will then be advised which treatment the particular bacteria is sensitive to, and will supply intra-mammary tubes for insertion into the teats immediately after milking. **Always milk a suspect last.**

Blown or Hoven

This is a condition of excessive gasses in the stomach, brought about by the goat consuming too large a quantity of succulent or frosted feed. It may also be a symptom, amongst others, of various other diseases which in themselves bring about paralysis of the digestive system.

Treatment

The distention will be particular*y* noticeable on the left side of the goat. It is necessary to massage the abdomen and to walk the animal around, not letting her lie about, whilst awaiting the dispersal of the gasses by the medicine. If the case is only a mild one, this takes the form of a heaped teaspoonful of indigestion powder, repeated after a couple of hours if necessary, together with a 2–4 oz dose of liquid (medicinal) paraffin, according to size. However, if the goat is badly blown, a proprietary drench of gaseous fluid, or the standard drench obtainable from a veterinarian's surgery, must be administered. In very acute cases a veterinary surgeon will puncture the flank to let out the gas.

Acetonaemia

This is a metabolic disturbance, first noticed by a pear-drop (or nail-varnish) smell of the milk, breath and urine. It may be mild and pass off after the goat has had her after-kidding worm dose, but if it is serious enough to cause loss of appetite, lethargy and a consequent drop in yield, veterinary advice must be sought or the goat may become comatose and die. Calcium borogluconate is used, and starchy foods, such as soaked sugar-beet pulp and mollassed crushed barley plus cortisone preparations.

Pregnancy Toxaemia

Known to be prevalent when there is a lack of greenstuff or fresh roots during pregnancy; also induced by lack of exercise at this time. The symptoms are not very definite, but any illness in the pregnant animal should be suspect and veterinary assistance called, as it may be necessary to bring on the kidding early (to save the goat) if treatment fails to alleviate the condition. Feed as above for acetonaemia.

Milk Fever

This is a misnomer, since there is no 'fever' but a sub-normal temperature.

Symptoms

The heavy milker, often producing her own body-weight in milk every 10 days, is under great strain. Even when presented with every food, mineral, trace element and so on, which she can possibly require to make the milk, she may still be unable to assimilate the necessary calcium.

At any time during her lactation (unlike the cow), even during the actual kidding, she may suddenly go off her feed, start breathing stertorously, appear slightly blown and lose interest in her surroundings, eventually falling into a state of complete coma in which, without immediate treatment, death will follow.

Treatment

A good cowman could help in an emergency, for he will be prepared, with a bottle of calcium borogluconate in stock, but the vet should be summoned at once if milk fever is suspected. He will inject direct into the vein if the goat is very ill. The goat must be propped up and not allowed to lie prone. Her recovery will be astonishingly rapid if the injection is in time, and she will soon be on her feet again and eating her hay.

A Repeat Likely

Animals prone to milk fever may suffer again in succeeding lactations, and pre-natal help should be sought to prevent this. Any colostrum surplus to her kid's requirements can with benefit be fed back to the goat.

Foot Rot

Even the best trimmed feet can have foot rot, which is much the same as this condition in sheep. Contaminated mud was thought to be the cause, but in the event no bacteria was found in tests on hoof parings and it seems likely that it is environmental. *Prevention* consists of avoiding the mud wherever possible, but if the goats just have to pass through it, a daily brushing over the sole and the sides of the hoof with Horse Guard hoof oil is of great benefit.

Treatment

Use terramycin aerosol spray on any lesions; to treat a bad case which is causing great pain and lameness, first scrub the foot thoroughly in warm water and Jeyes Fluid. Trim back, a little at a time, and gently examine any cavities. A foul-smelling black substance will probably be found. Wash this out, then pack the hole with cotton wool soaked in Horse Guard oil or Stockholm Tar. Cover with a home-made water-proof boot, easily shaped from a plastic bag and held in place with the iron oxide tape. The bag must be removed, the treatment freshly applied and a new bag fitted, at least every 24 hours, until the condition is alleviated.

Spots on the Udder (Cow or Goat Pox)

Worst in summer—must be kept dry and free from greasy udder salves. Fissan powder, which used to be most effective, is no longer available, so ask your veterinary surgeon to make you up something similar. Keep the bedding clean and dry.

Spots Under the Skin (on the neck and body)

This complaint commonly attacks Toggenburgs and British Toggenburgs—in other words, brown goats. It is often a rat-borne follicular mange, and will respond in time to washing with Cooper's No. 3 Gammexine Wash. The goat's coat is not dried off, but excessive liquid smoothed out and a brisk walk given to prevent chill. It does not seem to infect the same animal twice, but it does take a considerable time to clear up and may lose the animal the best part of a show season.

Prevention

Keep bedding clean and dry and eliminate rats. Persuade the cat to sleep by the hay rack.

Mud Fever

Symptoms

The heels and pasterns become hot, swollen, scurfy and the goat may be very lame.

Treatment

Paint the affected parts freely with Horse Guard Hoof oil—using a 2 in soft paint brush.

Contagious Dermatitis (of sheep) ('Orf')

This distressing complaint causes sores around the lips and nose, sometimes so bad that the animal cannot eat. The teats are also sore so the kids cause pain when suckling. It is extremely contagious and may even be caught whilst travelling in a lorry. *Orf* vaccine may be used, usually in autumn, where previous infection has been known. If a reddened appearance is noticed, with later a sore scab forming on nose and lips, call your veterinary surgeon for advice. Meanwhile, use Terramycin aerosol spray.

Lice

Lice may infest any goat, particularly in winter. A thorough dusting with louse powder (preferably out of doors) once a month, should be sufficient. If through bad weather the goats must be dusted and brushed off indoors, always remove all feeding utensils, water buckets and hay first.

Cloud-burst (False Conception)

Affects most mammals and is not clearly understood. Can affect mated and unmated animals and consists of a large amount of liquid causing the goat to appear to be in-kid. May take place at almost any time, from the date due to kid or the date the goat would have been due to kid if served at the last heat, to several weeks longer. The goat should not be served until any discharge has cleared up. Cloud-burst is often followed by repeated signs of oestrus, the goat coming in season every five days. Whilst this used to be regarded as hopeless in the goat (cows are large enough to manipulate manually) there is now treatment which can be very effective if carried out after the first two or three seasons.

Abortion

Occasionally a goat which has been badly frightened, or has squeezed through a narrow gateway, or has been fighting or has had acute indigestion, will abort. This is not to be confused with contagious abortion of cows, from which the goat does not suffer.

Treatment

Allow the best hay, plenty to drink, clean bed, and greenstuff and light mashes, as after a normal kidding. Be sure that the afterbirth is discharged whole, within a few hours of the abortion, and seek veterinary help if it is not.

Wounds

Cuts and scratches may be sprayed with the Terramycin aerosol spray. Keep away from the eyes. Use Animal Lintex for poulticing, as directions on the packet.

POISONOUS TO GOATS

Yew

Commonly a tree of churchyards, but sometimes seen hanging over a wall, when it can be dangerous, particularly in a gale.

Symptoms

Sudden collapse, scouring, colic, staggering gait.

Treatment

Call the veterinary surgeon, meanwhile give black coffee every two or three hours, and the mixture recommended below for rhododendron poisoning. Rug the patient and keep warm.

Rhododendron

Often a garden shrub, but should also be looked for on any woodland or country park land where it is intended to keep goats. The only successful treatment is very effective but must be given at once.

Symptoms

Much as above, for yew poisoning, but thought to be the only poison which actually makes goats vomit. They will throw up green slime and cud, will scour and eventually collapse.

Treatment

Rug warmly, prop up with deep straw. At once give, in the drenching bottle, a teaspoonful of bicarbonate of soda shaken up well in a good tablespoonful of melted lard. This *must* be pure lard. The dose may be repeated after two hours.

Laburnum

Said to have no known antidote, leaves, seeds and the bark, are poisonous, the seeds being the more deadly. The poisons are alkaloid.

Diagram 21-1—Yew

Diagram 21-2—Dogs Mercury

Spindle has been consumed without harm, but is listed as toxic.

Privet and **laurel** both contain certain amount of glycoside and although not deadly, should be avoided.

Dog's Mercury *(Mercurialis perennis)* growing, as it does in early spring, may be sampled because there is not much about.

White Bryony grows in the hedges—the roots and berries being the most dangerous parts.

Rhubarb leaves are very poisonous, and should always be burnt or buried in the compost heap.

Ragwort All parts are poisonous, but this plant is most dangerous when it occurs in hay. Controlled by the Injurious Weed Regulations, but with the loss of rabbits and Cinnabar moths (which are its natural controls) the ragwort is spreading, particularly after drought. The poison is accumulative and destroys the liver.

Water Dropwort is dangerous in early spring, before the grass grows, as it will be eaten in large quantities. Although it is the white rather parsnip-like root which is the most toxic part, the leaves cause scouring and loss of appetite and consequently loss of milk.

Cowbane is very like water dropwort and grows in the same damp conditions.

Hemlock and **water hemlock** *(Denanthe crocata)* are both poisonous and should be avoided if possible.

Accumulative Poisons

Bracken, particularly when dried for bedding, and potato peel or green chat potatoes, uncooked, may not give rise to immediate symptoms, but may prove toxic after some time. Toxic sprays may also affect animals; also rat poisons, lead and copper containers, and cereals and protein cakes which are stale, damp or mouldy. Mouldy hay would not normally be eaten by goats, but, if taken, would be dangerous.

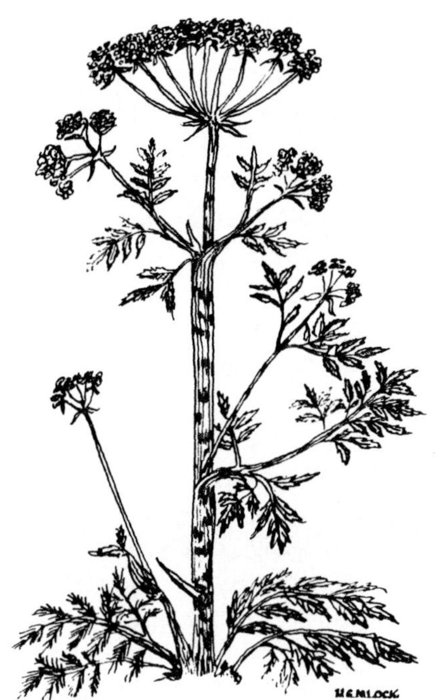

Diagram 21-3—Hemlock

Diagram 21-4—Water Hemlock

General Rule for Suspected Poisoning

Try to find cause and inform a veterinary surgeon. Give black coffee or strong tea, after 4 oz of medicinal liquid paraffin, rug the patient, and prop up with deep straw, keeping as warm as possible. Milk is often helpful, if the goat will drink it, and glucose may be added with advantage.

A stall-fed animal taken out for exercise, and one which is short of greenstuff, is much more vulnerable to any poison than are goats on free range with ample grazing and/or browsing.

A great many household goatkeepers used to take advantage of wide grass road verges, cutting this for hay, and using some of it green. However, these days it is seldom worth the risk, for it may have been sprayed with a weed killer so toxic that even breathing a few whiffs of it can make the sprayer ill.

A case of death in young foals should be mentioned, too, for this is a hazard that nobody would think of, until a tragedy happens. A small child, finding some little pellets the size and shape and colour of calf pencils, took a handful from a jar in his father's greenhouse and offered them to a foal a few weeks old, which ate them readily from his hand and died an agonising death. They were slug killing pellets.

A SPECIAL NOTE

There are many more plants and shrubs listed as poisonous, but on the whole goats are very careful feeders. Those mentioned are the most likely to be troublesome and consequently a special note should be made if they are seen growing within reach of goats.

CHAPTER 22

PASTURE AND CROP GROWING

GRAZING

There is no breed of goat which will not learn to graze well, if the pasture offered is really good and at the right stage of growth. Goats do not like long grass, or very short grass; the perfect length is about 4 in, and considerable skill is required to keep it at the right length. Here the electric fencer comes into its own, because the goats can be strip-grazed and moved on every few days, if required, being followed up by a pony or a cow, after which the first strip is rested and allowed to grow again.

FERTILIZERS AFTER SOIL TESTS

Having the soil tested and any deficiencies corrected is not only sound policy for the animal's health and milk yields, it is also going to make certain of a good 'take' and eventually a good crop from your new ley. To leave old rough tired grass down is folly. It is fairly easy nowadays to hire a contractor if you do not have the tractor or horse and implements for ploughing.

When a new grass ley is to be sown, early autumn is the best time for ploughing. The seed mixture should be sown in March or early April, and some suggested mixtures suitable for goats are given.

FIRST GRAZING

The new ley in a good year will be ready to graze for a few days after about six to eight weeks, and the small feet and light weight of goats help to consolidate and do no harm, while the biting (not tearing) action of the goat helps to thicken the sward.

The Author prefers to turn out the goats early rather than late, when there is not too much lush young growth. Thus they become accustomed gradually as it grows and they do not become blown. Always feed hay before turning them out and be ready to move them off if they seem to have had enough.

At certain times, it is desirable and quite practical to use a renovating mixture, without ploughing up the field; as, for example, after severe drought has killed off some of the finer grasses and left gaps. The Author does not like the use of chemical killers.

Each spring, the pasture should be rolled and chain-harrowed, the fields taking it in turn to have spread what manure is available. If the available goat manure will not go round both the pasture and the crops and vegetable garden, some proprietary grass fertiliser should be used, particularly on a field to be cut for hay, but do not turn the goats on to this until rain has melted it well.

HAY

Hay can be cut with a small mechanical cutter and hand-made for just two or three goats—a few home-made tripods will be very helpful for drying it. After it is dry it may be baled by using the home-baling plastic bin shown in the diagram, and pressed down and tied, making storage much easier.

Small quantities of other crops cut with the intention of using green, if spare, can also be made into excellent hay. Lucerne, giant red cow clover, oats and tares (vetches) all make lovely hay if dried on the tripods. Nettle hay is very popular.

KALES

Of the kale varieties, Canson is first choice; it grows into a huge plant if well manured and spaced about 3 ft apart. First the lower leaves are plucked, then the plants can be beheaded and, when sprouted again, grazed if required. One large plant pulled when fully grown will feed three or four goats for one feed. It should be sown on well-manured and worked ground in late March or early April to make big plants before winter.

Marrow stem kale is best broadcast, when it can be scythed for use; if drilled and thinned, the stem becomes too hard for goats and usually has to be chopped.

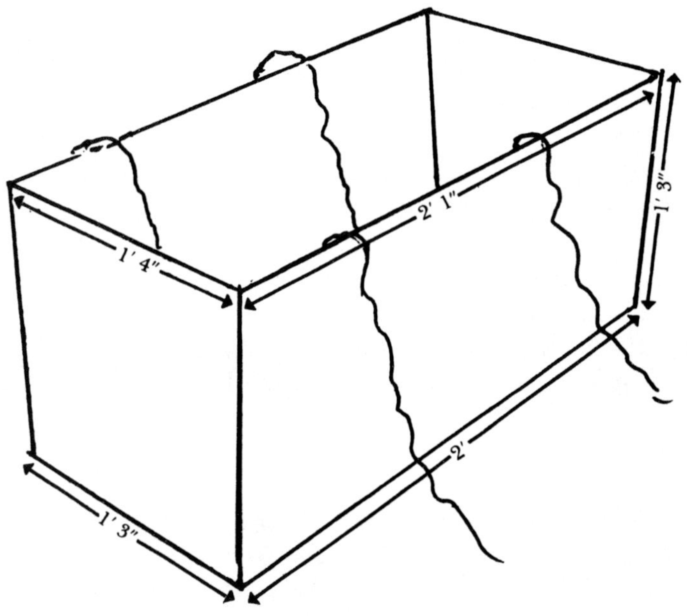

Diagram 22-1—Container Used for Making Home Made Hay Bales

Hungry Gap kale is a smaller plant, useful, as its name implies, for feeding in March and early April. It should be sown in June and is best drilled, the plants becoming about 18 in high, as opposed to the 4 ft high Canson.

The tips of all the kales make good eating for the household and cutting them out only serves to make the plant bush out.

There is another perennial kale plant which is useful for a corner of the garden that will not be disturbed. It is everlasting, and is propagated by cuttings. Pieces can be broken off and fed at any time.

CABBAGES

A few rows of Flatpol cabbages are a very useful standby, because in very wet weather they are much easier to dry off before feeding—they can be shaken and hung upside down for a while. They are so big when well grown, that one cabbage makes a feed for two or three goats, but like all the brassicas they require adequate manure and lime, and usually some artificial manures as well.

MAIZE

Fodder maize is a good crop for feeding at midday in the height of summer when the goats may have to come in, or for taking to shows, but first make sure that your particular animals like it—only one of the Author's goats will eat it.

The seed is sown in May, in rows 16 in apart; the depth to sow is about 3 in.

OATS AND TARES

This is a favourite with all goats and a good cropping mixture which makes a well-balanced feed towards milk production.

The seed is broadcast sown in early spring for summer use, and the winter variety can be sown in October for early spring use.

It is cut when the oats are in the milky stage—once it has reached the ripe hard stage it should not be fed unless thrashed and bruised. To make hay of any small amounts not required for feeding it must be tripoded and completely dried.

LUCERNE (Alfalfa)

A strip of Lucerne is an extremely valuable crop in time of drought as the roots go down so deep that it is scarcely affected. Lime is absolutely essential and so, also, is nodule-forming bacteria culture, which is usually sold with the seed.

The seed is inoculated with the culture, according to maker's instructions, then drilled at the rate of 30 lb per acre, from April till July. The drills should be 8 in apart.

As soon as the seed has germinated, the rows should be hoed down regularly to keep the weeds from taking over. If it can be kept free of weeds, lucerne can be cut about 10 weeks after sowing for the first time and, thereafter, three or four times every year for about five years.

For making hay in the small amounts, the same rules apply as for oats and tares; the tripods will be needed and the hay must be really dry before baling.

ROOT CROPS

In order of general preference, otofte fodder beet, mangolds, carrots and swedes, are the roots most usually grown for goats.

Fodder beet is treated in much the same manner as mangolds, but as it is not such a large root as the latter, it need not be allotted quite so much space per plant. It is sown in April or May, and can be pulled and clamped as are mangolds, but the leaves are best ploughed back into the soil, not fed to the goats.

Mangolds contain more water than fodder beet and must not be fed at all until they have ripened in the clamp after Christmas. They can then be scrubbed and cut in half. The cut side uppermost when placed in a bucket will be vigorously eaten and they give something to relieve boredom on a wet and cold winter's day, when the goats are confined to their house. Male goats must *not* be given mangolds.

Carrots and swedes are very good food value, but do not grow in every part of the country; varieties likely to be useful will be found in the seedsmen's catalogue of the individual locality. Times for sowing to the best advantage are likely to vary too. June is about the most usual month for swede sowing.

PROTECTION

All the brassica and root crop seeds are usually sold already treated against the flea beetle—which attacks before the leaves have passed the first stage and can destroy the crop.

For cutting up roots, particularly fodder beet and swedes, a home-made root-cutter such as that illustrated in Chapter 12 is a great help.

A clamp for keeping roots frost-free through the winter, should be made in a sheltered spot by heaping on hedgings, bracken, heather (as available) or straw (failing these) and covering with plenty of earth.

TO DRY HAY ON TRIPODS

The grass cut for hay should be well wilted and never wet, either from rain or dew, when it is put on the tripods.

First pick up a forkful of wilted grass and place it on the cross rails at one corner of the tripod. The next forkful is placed adjacent to it on the side rail—continue working round the tripod until the starting point is again reached. Continue round and round to the top of the tripod, finishing with a good big forkful right on the top.

Next, rake down the sides with your fork so that the loose hay is pulled off and all the ends are hanging downwards. This helps to shed rain. The whole now looks rather like a well-groomed giant toadstool and, if one creeps under it, the structure will be found to be hollow. This allows the air to circulate through the hay, drying it and preventing moulds.

In a good season the hay is ready for storage in about 10 days. It takes considerably longer in a wet summer, but it will not be spoiled. In any case, there is no hurry to get tripoded hay into the barn as it will remain in perfect condition for several weeks. High humidity districts in which giant red clovers, lucerne, oats and tares, and other thick stemmed hays which cannot be made on the ground, will prove no handicap with the use of tripods.

Making tripods for Haymaking

Each tripod consists of six poles of nut, ash, willow or holly, etc. Three of these are about 9–10 ft long and tapering from about 6 in in diameter to about 3 in in

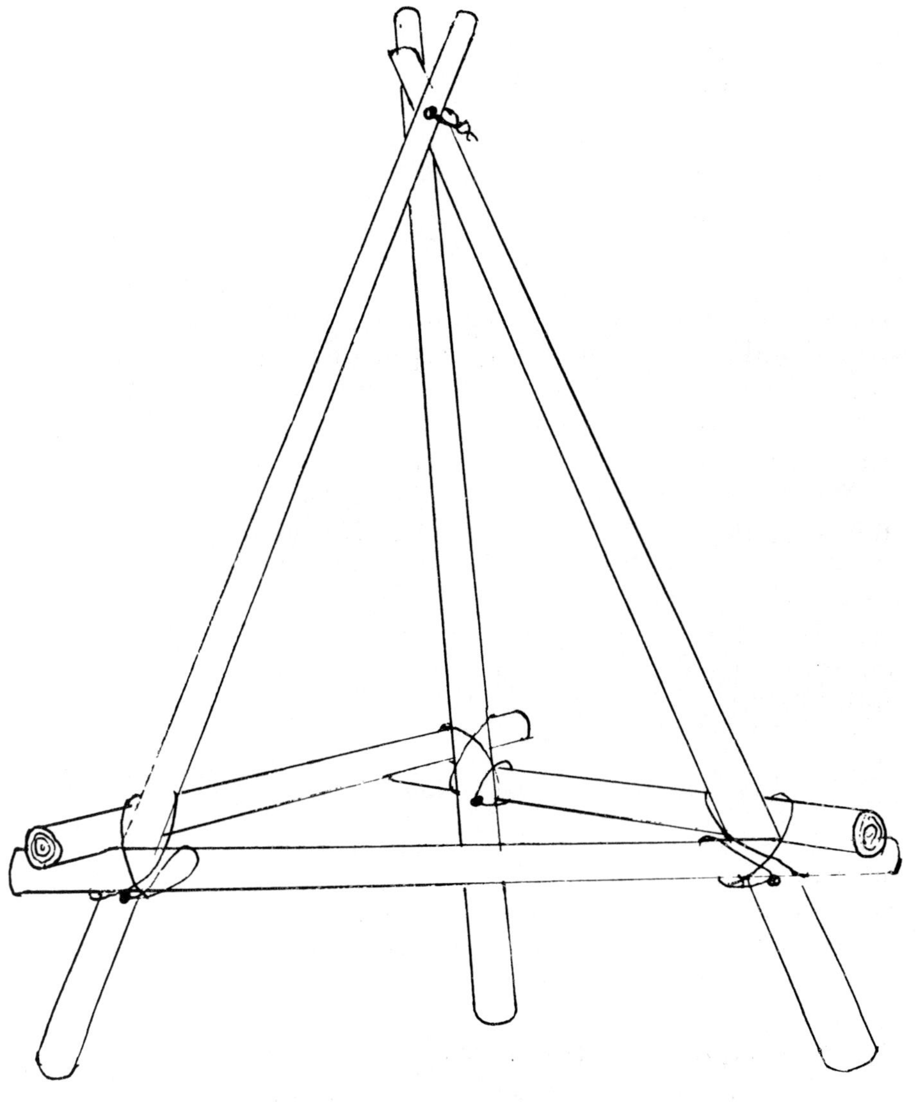

Diagram 22-2—Tripod for Haymaking

diameter at the top. The side rails should be about 6–9 ft long and tapering from about 4 in in diameter.

Drill one ½ in hole in each of the long poles, about 1 ft from the top. Drill another hole about 2 ft from the thick end.

A stout wire is passed through the top holes and the ends fastened securely together, leaving enough slack to allow the poles to be extended at the base, to about 6 ft apart.

The side rails are then wired on, using the lower holes to stop them from sliding down.

This makes a very stable structure. The tripods can be dismantled after use and if stored in a dry place will last years.

INSTRUCTIONS FOR HOME-BALING HAY

Take a strong container about 2 ft long, about 1 ft 3 in wide and deep. A strong wooden box will do, or a fibreglass water tank, which must be one without an inside rim.

Lay two strings across the box, as shown in the diagram, and fill it with your well made hay, treading it down as you fill the box.

When no more will go in, tie the strings across very tightly and then lift the bale out ready for storage. A container that is very slightly narrower at the bottom than at the top makes lifting out easier. *See* Diagram 22-1.

SOWING PASTURE

A Suitable Herbal Strip for Goats

1·50 kg Aberystwyth S23 perennial ryegrass (Pasture Strain).
1·25 kg Canadian altaswede late flowering red clover.
1·75 kg chickory.
0·25 kg ribgrass.
1·00 kg sheep's parsley.
0·10 kg yarrow.
0·50 kg Grasslands Huia white clover.
6·35 kg per acre.

For Re-seeding Heavy Soil

2·00 kg Lema Italian ryegrass.
5·50 kg Aberystwyth S24 perennial ryegrass.
1·50 kg Combi Italian ryegrass.

Figure 12—Feral goat from the College Valley, Cheviot Hills, Northumberland

(Now on the farm of Mr and Mrs K. Briggs in Worcestershire (photo Ken Briggs))

1·75 kg Melle perennial ryegrass.
1·75 kg Clinax Timothy.
1·00 kg Kampe 11 (Timothy)
1·00 kg Grasslands (Kahu Timothy).
1·00 kg Canadian Altaswede late flowering red clover.
0·50 kg Superfine Canadian Alsyke.
0·50 kg Grasslands Huia white clover.
17·25 kg per acre.

For Heavy Cuts of Hay

2·00 kg Dilana Tetraploid Italian ryegrass.
4·50 kg Aberystwyth S24 perennial ryegrass (hay strain).
2·50 kg Hungaropoly Tetraploid broad leaved red clover
9·00 kg per acre.

The figures given are intended as a guide only; obviously the costs will vary from year to year.

INDEX

Note: The names of goats are printed in italic, but suffixes and prefixes awarded have been omitted. Numbers in italic refer to the page numbers of illustrations.

Symbols